The Unique Animal

The Unique Animal

the origin, nature & consequences of Human Intelligence

by
Don D. Davis

I

 Prytaneum Press

London & New York

First published, 1981
Copyright © Don D. Davis, 1981

ISBN   0 907152 02 3 (hardback)
           0 907152 01 5 (paperback)

Prytaneum Press
121 Bouverie Road
London, N16 OAA
England

Prytaneum Press
16 Fair Way
Poughkeepsie, N. Y. 12603
U. S. A.

British Library Cataloguing in Publication Data

Davis, Don D.
      The Unique Animal.
      1. Intellect
      I. Title
      153.9   BF431

      ISBN 0-907152-02-3

Printed in Great Britain by

Lithosphere Printing Co-operative Ltd.
467 Caledonian Road
London, N7 9BE

Set in Bold Face II

To Virginia

What does a fish know about the water in which he swims all his life?

Albert Einstein

Preface

The hard line or radical school in the field of learning psychology holds that any theory should be viewed with extreme suspicion--if a theory is to exist at all, it should be "built up" from the results of numerous experimental journal articles. No hypothesis or theory should be put forth, it is claimed, unless it can be backed up on every point with several references from experimental studies. It is believed that no theory should make statements beyond what the currently known data support. Thus a scientific theory should be slowly constructed in a way analogous to a mason building a wall--one brick at a time. The acknowledged leader of this behaviouristic school of psychology, B. F. Skinner, wrote an article entitled "Are Theories of Learning Necessary?" (1950). (He concluded that they are not, and many behaviourists still cling to this view). This simplistic view of science is not only historically inaccurate but also completely unworkable. Scientists in other fields have long abandoned this naive view. The history of science shows a continuous series of successive hypotheses and theories put forth to explain relevant events in all areas of study. Without hypotheses and theories, science could not exist.

Of course, the first prerequisite of a good scientific theory is that at least part of the theory should be testable and falsifiable at least in principle. Probably the reason why many of today's learning psychologists are dubious of theories is because of the difficulties that were posed by the elaborate theories constructed by Freud, Jung and others in the first part of this century. These early theories proved to be difficult, if not impossible, to test, even in part. After having a bad experience with these first theories, behaviouristic psychologists have become very cautious and sceptical of any comprehensive theory. This is unfortunate, because they have in effect thrown the baby out with the bath water. The fact that the first theories in psychology were poorly formulated from a scientific view does not justify condemning all theories. Theories themselves should not have been made suspect, only ill-formulated theories.

Another important aspect of a good scientific theory that is usually not clearly recognized is that a theory should not only offer an explanation by proposing a relationship between events, but it should also point beyond the existing facts and make at least some statements that have not as yet been experimentally verified. A theory that did not state something new and previously unrecognized about the world would not be much of a theory. Almost by definition, a theory should propose something about nature beyond what the current collection of facts makes evident.

To take some of the better known examples, the theories of Copernicus, Darwin, and Einstein all made some statements about nature that could not be confirmed from the current knowledge. All made statements or predictions beyond the evidence of the day. As a result, all these theories stimulated research on the relevant events, the importance of which had not been previously realized. Thus it was only after Einstein published his theory of relativity, which predicted that gravity could bend light waves, that physicists first thought to check the light from stars whose light rays passed near the sun during a solar eclipse. These observations supported Einstein's prediction. Thus a good theory should point the direction for experimenters to study by making predictions beyond the existing evidence. In sum, theories should go beyond the present data, but they should also be testable. This double requirement might have something to do with why good theories are so few and far between.

Unfortunately, experimenters without a theory to point the direction, tend to spend a lot of time groping about, conducting many trivial and unimportant experiments and only occasionally stumbling upon some important discovery. Scientific experimenters without a theory are somewhat like a blind man swatting at phantoms in the night. They only occasionally hit something, and then only by sheer chance. True, this "blind experimenting" is often necessary, but scientists should never deceive themselves that it is the ideal state of scientific investigation. Any falsifiable theory is better than no theory at all. Or as Francis Bacon remarked, truth will sooner come out of error than out of confusion.

Actually, scientists who claim they are collecting raw experimental facts without the aid of a hypothesis or theory are deceiving themselves. Any experiment must have at least a tentative, rudimentary hypothesis that there might exist some type of relationship between the variables being manipulated. The very fact that they select a few variables for study out of the literally hundreds of thousands that are known to them proves that a primitive hypothesis has already been formed. A hypothesis that is little more than a semi-conscious hunch, is in no way superior or more scientific than a hypothesis that has been carefully thought-out and consciously analyzed. On the contrary, I suggest that the opposite is true.

What follows is a theory of several important aspects of human nature. I did not conduct a single laboratory experiment in the preparation of this book, but of course many experimental results conducted by others are incorporated within the theory. Also many anthropological, philosophical and historical studies are cited. The theory is rather the result of many hours of thinking and reading. Although I conducted no experiments, I conducted thousands of "thought experiments", trying to fit the uncountable pieces of evidence into a meaningful picture that fits the facts. The important point however is that this theory is testable and falsifiable on at least several core points. How a scientific theory is formulated is not important as long as it is possible to hold the theory up to empirical testing.

The theory makes three main new points:

1) That man is capable of a certain type of learning of which no other animal is capable.
2) That as a result of this unique type of learning, humans are the only animals that anticipate as a result of experience which events are likely to occur more than one minute into the future.
3) And that the extent of this unique human learning in each society has a profound effect on that society's laws, morals and customs.

DDD

London, 1981

Acknowledgements

I am most grateful to all the following people who read the manuscript of this book and made suggestions for improvement: Virginia Styman, Alison Styman, Tommy & P.A. Cole, Al Erickson, Marshall Davis, Peter Sutherland, PP-C, Lucy Mair, Michael Simpson, Anthony Dickinson, R.W. Ashby and G. Dunk.

I also wish to thank Herbert van Thal, J. G. O'Kane and Jim Pennington for their helpful suggestions concerning publishers and publishing.

Also, special thanks to Floretta Davis, without whom, none of this would have been possible.

And last, thanks to T. B. Davis for giving me the financial resources necessary to complete this project.

Contents

PART ONE

Chapter 1

The Intelligent Animal

> Man is an animal who thinks.
> Lucius Annaeus Seneca

Many is the child that has looked deep into the eyes of his pet dog or cat and wondered, if for only a moment, about the nature of the barrier that separates humans from other animals. Throughout human history many adults have also thought a great deal about this difference between man and the beast. Today we may smile to ourselves at the naivety of the above statement by the Roman statesman Seneca. But for all our great advances in knowledge in recent centuries, we are still woefully ignorant of the nature of our own intelligence and exactly how our intelligence differs from that of other animals.

It is obvious, not only to the scientists and philosophers, but also to the common man that humans are in some way smarter or more intelligent than any other animal. The contrast between humans and non-humans is quite apparent, but the exact nature of the difference in mental ability and the ramifications of this difference are still to this day very much a matter of speculation and dispute. Men agree that humans are more intelligent than other animals, but there is no agreement as to the specific nature of this intelligence.

Perhaps one of the reasons why we have had such difficulties in analyzing our own intellectual abilities is the impossibility of getting outside ourselves. We are trapped in our own bodies. We always see things only through human eyes and can think only with our human brains. One often hears from world travellers that they were never able to fully understand their own country and its values until they had travelled abroad and visited other countries with somewhat different values. By getting outside their own country and by experiencing different cultures, these people are able not only to learn about these foreign countries, but also, by comparison, to understand their own culture much better. Mankind then, must be somewhat like the person who would dearly like to visit foreign countries, but is never able to do so. We are never able to experience what other animals are experiencing. We can never really know what is going on inside their heads.

However, even the person who is unable to leave his own country has some information about foreign cultures available to him. He is able to read books about other countries or even talk directly with foreigners (assuming they speak a common language) or nowadays see films from abroad. But as humans, we are never able to leave our human bodies and experience how other animals are experiencing this same world. And what is more, unlike the foreigners, animals do not talk to us and tell us about it.

Perhaps our task would be a great deal easier if somehow one of us could become another animal for a few days and then return to tell us what it was like, or if another animal could acquire the gift of speech and tell us about its former life. But these methods are all but impossible. We must, it seems, seek the answers elsewhere.

Science has often sought to explore mysteries of nature that were beyond the direct observation of human eyes. In problems of the infamous "black box" type, science has often excelled in providing useful proposals for the understanding of these mysteries. Scientists have often proposed the existence of objects that could not be directly observed at that time. Some examples of these proposals are molecules, bacteria and genes. By carefully observing what goes in and what comes out of one of these "black boxes", scientists can sometimes hypothesize a good approximation of what may very well be happening inside the box, even though they are not actually able to observe what is happening on the inside.

Perhaps science can even yet do the same for the problem of human and animal intelligence. The animals cannot tell us what is happening inside their heads, and we cannot experience directly what they are experiencing, but by carefully observing the differences between human behaviour and animal behaviour, perhaps a scientific theory can be formulated to account for some of the differences.

DEGREE VERSUS KIND

Starting with Aristotle, many philosophers and scientists have pondered the question of the nature of this difference between man and the other animals. The controversy over the centuries has tended to centre around two main points. The first of these points concerns the amount of the difference of intelligence between man and the other animals. That is, whether this difference is a matter of degree or rather, a matter of kind. Even to this day it is possible to find equally qualified scientists arguing over whether man is more intelligent only by degrees from the other animals or whether man's mental activity is of a different kind from the rest of the animal kingdom.

Aristotle in his *Historia Animalium* was one of the first to defend the "degree" side of the argument. He contended that the dog, the elephant and other animals approached the mental level of the human child. Charles Darwin (1871) in more modern times tended to agree with the view of Aristotle when he stated that "there is no fundamental difference between man and the higher mammals in their mental faculties", that this difference consists solely of man's "almost infinitely larger power of associating together the most diversified sounds and ideas. . . the mental powers of higher animals do not differ in kind, though greatly in degree from the corresponding powers of man".

In this century, Edward L. Thorndike, the first man to study animal intelligence in the laboratory, was also of this "degree" opinion. He denied not only any unique intellectual difference between man and the other animals, but also any uniqueness anywhere in the hierarchy of the animal kingdom. He believed that intellectual differences between animal species were only differences of degree. "All the vertebrates apparently have the ability to learn. The general pattern and features of the learning are extraordinarily alike over almost the entire range. Molluscs and arthropods, fishes, amphibians, reptiles, birds, and mammals manifest fundamentally the same process of learning" (1931, p. 163).

Defending the other side of the dispute, Descartes (1637) argued in favour of the differences-of-kind position. It is "not only that the brutes have less reason than man, but that they have none at all". John Locke (1706) also saw that "the power of abstracting is not at all in them /other animals/ and that the having of general ideas is that which puts a perfect distinction between man and brutes".

The experimental methods of modern science have by no means settled the issue. Contradicting Thorndike's hypothesis of difference of only degree in the hierarchy of animals, Bitterman (1965) has found in his experiments that there is a discontinuity in animal learning along the evolutionary scale. For example, in his habit-reversal experiments, Bitterman found that the higher animals were able to improve their performance on this type of problem, whereas the lower animals were not. In these experiments, animals were rewarded for choosing the alternative A rather than B until a preference for A had been established, then B was rewarded instead of A. After the preference for B had been established, the reward was switched back to A and so forth. After the habit was established, the reward was reversed. Animals such as the rat and monkey were able to show a steady improvement in switching back and forth. However, the fish in contrast, were never able to show any improvement whatsoever, later reversals being accomplished no sooner than the earlier ones.

The most conspicuous modern defender of the idea that man is different in kind from all other animals is Leslie White. He believes that human language is different in kind from other types of animal communication. "Man differs from the apes, and indeed all other living creatures so far as we know, in that he is capable of symbolic behaviour. With words man creates a new world, a world of ideas and philosophies" (1949, p. 46).

PROPOSED UNIQUE ABILITIES

In addition to this dispute of the "degree versus kind" of human intelligence over animal intelligence, there has also been superimposed throughout the years the question of what, if any, specific abilities man possesses that other animals do not. For example, it has been proposed at one time or another that man is the only animal able to use tools, or to use abstract thought or to use language and so forth. The same men who believed that humans are different in kind in intelligence have raised these proposed specific differences to support their claim of human uniqueness, and the men on the degree side have done their best to show that these specific differences are not actually a difference of kind but rather only a difference of degree.

The people on the "kind" side have proposed specific abilities which, although humans are easily able to perform, other animals, it has been claimed, are *never* capable of doing. Their opponents have attempted to show that other animals are at least occasionally capable of these activities. Once it had been clearly demonstrated that some of the other animals were capable of performing one of these proposed unique activies *even on a simple or primitive level*, it has generally been agreed that this proposed difference of kind was disproven and was actually only a difference of degree. It is important to remember how these terms "kind" and "degree" have been used in this area in the past as this distinction will be discussed again in Chapter 4.

The following is a brief summary of some of the various abilities that have been put forward at one time or another to mark man apart from the rest of the animal kingdom. Even though some of these supposedly distinctive abilities have now been discounted, each, in its day, had its defenders and its detractors.

I have only listed those proposed abilities that science has been able to examine at least to some extent. Saying that man is the only "rational" animal may sound promising at first, but no one has ever been able to define exactly what rational behaviour is (as opposed to non-rational behaviour). Even if we assume that humans are indeed "rational", how is it possible to show that other animals are not rational? Much the same is true for the idea that man is the only animal that has "self-transcendence". Even assuming that man does have this ability, how can we test other animals to see whether they do or do not have this "self-transcendence"? For all I know, the cat sitting on the fence outside my window as I write this may be transcending himself at this moment, but I have no way of knowing one way or the other, and until someone is able to define self-transcendence in such a way that we are able to test animals and ourselves for the possession of this presently ethereal ability, we should leave the question open and admit our ignorance.

PHYSIOLOGICAL DIFFERENCES Before considering the various proposed mental abilities between man and the other animals, however, it might be useful to first examine the physiological differences between humans and the other animals. When examining these physiological differences--as well as the various proposed mental abilities--it will be of particular interest to look closely at the differences between man and his nearest relatives--the great apes. Various studies have convinced most scientists that next to man, the most intelligent animals are the apes. In studying the mental differences between humans and the other animals therefore it is essential that particular attention be paid to the differences between man and the apes.

At one time it was argued that the physiological differences between humans and apes were relatively small, so that in fact the apes (gibbon, orang-utan, gorilla and chimpanzee) differed more from each other than they did from Homo sapiens. However this contention is no longer accepted. Today it is generally agreed that man differs more from the apes than they differ from each other.

Schultz (1936) has listed thirty differences peculiar to man from all the apes. Some of the better known of these differences are: the greatest weight at birth in relation to adult body weight, the largest relative brain size, complete bipedal walk and erect posture, the greatest reduction in density of hair--except on the scalp, the occurrence of wavy and curly hair, the complete lack of a penis bone, the

highest total number of vertebrae, the unique structure of the kidney,
by far the lowest shoulders and the lowest placed nipples, by far the
longest thumb in proportion to the length of the hand, the equality of
the sexes in regard to the size of the canine teeth and so on.

For our purposes however, the second difference listed above--
the largest relative brain size--is by far the most interesting and bears
looking at more closely. The human's brain case is by far the largest
in capacity, having roughly three times the capacity of that of the
largest brained ape. Our skull averages about 1,450 cubic centimetres
which compares to about 500cc for the gorilla, 404cc for the chimp,
395cc for the orang-utan and 128cc for the gibbon.

Not only do we have a large brain, but when compared to the brains
of lower animals, we find that most of the growth of our brain has been
in one portion of it--the cerebrum. The massive cerebrum dominates
the appearance of the human brain. It is formed in two hemispheres
perched on top of the rest of the brain. In the lower mammals, the
cerebrum is small and is mostly devoted to the sense of smell. In the
higher mammals, the cerebrum becomes larger and larger until in
man in constitutes over 90% of our brain volume.

The surface of the cerebrum is known as the cortex. The cortex
is three to four millimetres thick and wrapped in wrinkled folds around
the two cerebral hemispheres. In humans the cerebrum still controls
sensory perception, and certain areas of the cortex can be assigned to
the various senses. In the lower animals, these sensory and motor
areas are bunched together with little room between them. However
the primates have developed *association areas* of the cortex surrounding
the sensory areas. These association areas do not have any direct
projections outside the cortex. They relate to the outside world only
after other portions of the brain have already processed the information.

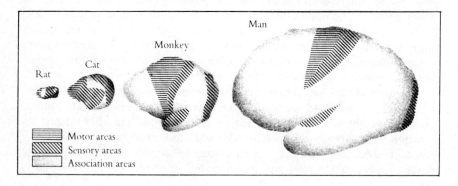

Man

Monkey

Cat

Rat

▭ Motor areas
▨ Sensory areas
▢ Association areas

These association areas are by far the most extensive in man. In the diagram, note the increases in both the absolute and the relative size of the association cortex in four mammals. In humans, these areas include the prefrontal lobe and regions of the occipital, temporal and parietal lobes. The precise function of these association areas is difficult to determine, but they are clearly acting upon information which has already received sophisticated processing.

Since these association areas are largest in humans and since we have the most highly developed intelligence, there is reason for assuming that there must be some sort of relationship between our intelligence and these association areas. If man does have some unique intellectual ability, it may very well be the association areas of the cortex that provide the seat for it.

TOOL USING AND MAKING For many years it was generally believed that man was the only animal able to use tools. We now know however, that there are many other animals that have this ability.

There are several birds that are able to use a tool to help them acquire their food. The Galapagos woodpecker finch can use a cactus spine or twig to pry out insects from crevices in bark. This finch even holds the twig under one foot while eating and will change ends of the stick when one end becomes blunted. In addition, the Egyptian vulture uses a rock to help it break open ostrich eggs. And it has recently been shown that blue jays are able to crumple up a piece of newspaper and use it as an extention of their beak to rake distant food pellets into the cage.

But birds are not the only lower animals able to use tools. Californian sea otters regularly use stones to smash shellfish. And of course, non-human tool use becomes most advanced in the baboons and especially in the chimpanzees. Chimps appear to be unique in their ability to use different tools for different purposes.

Kohler (1927) showed in his extensive experiments with captive chimpanzees that they are able to use a wide range of tools quite spontaneously. They are able to use sticks to draw in food beyond arm's reach into the cage. Some chimps are even able to use a short stick to draw a longer stick into the cage and then use the longer stick to draw in the distant food. Exceptionally clever chimps are able to stack as many as five boxes under a banana hung from the ceiling and climb the boxes to obtain the food. They can use sticks to pry open box lids and to dig roots from the ground. They use straws to draw in columns of ants to eat the insects and so on.

It is quite clear then that many other animals are capable of using tools. It is thought that Benjamin Franklin first suggested that man is the only tool-*making* animal. This assertion of man's unique tool making ability also has fallen recently after careful observation of chimps in the wild by Jane van Lawick-Goodall (1971). These chimps were able not only to use objects as tools, but also to make modifications to some objects demonstrating the crude beginnings of tool-making. They used grass stems to draw out termites from mounds. The chimps would strip the leaves off a stem to make a suitable tool. As the stems became bent, they would bite off the ends or use the other ends or discard them in favour of new ones. They would sometimes travel up to 15 metres from the mound to find firm stems and also collect three or four stems at once, return to the termite nest and lay the spares on the ground until they needed them. These wild chimps were also able to make another tool--a home made sponge. They chewed leaves, making them more absorbent and used them to draw water up from tree hollows.

So man is not the only tool user nor the only tool maker. Although several other animals use tools, it seems that only the chimp among the lower animals is able not only to use a variety of tools for different purposes, but also to show the crude beginnings of tool-making. However, in this area, it seems that man still can claim to be the only animal that uses one tool to make another tool. For example, even the chimp has been unable to use a stone hand axe to split a piece of wood into splinters which it needed to obtain food from a narrow pipe. Khroustov (1964) found that a chimp could split off pieces of soft wood with its teeth for this purpose. But even though shown many times how to use the hand axe on tougher wood, it was never able to use this axe to produce splinters.

Before moving on to the next proposed unique ability of man, it might be of interest to ask what we know of man's tool using and making past. When did our ancestors first start to use tools? Unfortunately, we are unlikely ever to learn when this happened. Our ancestors' first tools were most certainly common stones, shells, sticks, antler and bone. Since tool *use* seldom leaves any archaeologically identifiable remains, the date of our first tool use is likely to remain a matter of dispute. Even if an archaeologist were lucky enough to see a stone that had actually been used as a tool countless centuries ago by early man, this stone would most likely look like any other stone, and the archaeologist would not be able to realize the importance of it.

However, we do have some indirect evidence of man's tool using

past. In a cave along with early man-ape fossils, Dart (1959) found
numerous baboon skulls which showed signs of having been fractured
open. The nature of these fractures leads Dart to conclude that these
baboons had been killed by blows from a club. Also found in the cave
were large fore-limb bones of animals that match the size and shape of
the baboon factures. In other words, early tool using man may well
have used these bones as clubs to kill baboons. Not all archaeologists
are convinced, however, that there is enough evidence in this collection
of bones to draw this conclusion.

   Our first attempts at tool-making are unlikely to have been preserved. Like today's chimps, we probably started our tool-making
activities with sticks and stems and other easily manipulated materials.
This type of material usually decays within a short time. It is only
when tool-making man turned his attention towards stone that we are
provided with a certain archaeological record of his activities. The
first stone tools we have so far found are the crude "chipped pebbles"
made at least a million years ago and perhaps as much as two million
years ago in the lower pleistocene epoch.

   These stone pebble tools must have been made by using another
stone to chip off the desired bits, that is using one tool to produce another tool. This is the feat that modern chimps can still not accomplish. Our first solid evidence of early man's tool activities is not of
his first tool use or even his first tool-making, but rather of the more
advanced stage of using one tool to make another tool.

ABSTRACTION Since the time of the ancient Greeks, men have been
asserting that humans have the unique ability to "think abstractly", or
to form general concepts. Although clearly not as vague as the term
"rational", the problem of defining it and of demonstrating experimentally that only humans have this ability has so far been difficult and
inconclusive. The fullest account of the research in this area has been
prepared by Pikas (1966).

   Aristotle in *De Anima* defined abstraction as disregarding the par-
ticulars in order to extract what is in common. Wundt (1894) made a
distinction between two types of abstraction. He defined "isolating ab-
traction" as merely discriminating activity and "generalizing abstraction" as "neglecting the attributes . . . of a number of objects or facts
. . . in order to retain certain of them that are common to the whole
group and raise them to the status of characteristics constituting a
general concept. "

   In more recent times, Goldstein and his colleague (1941) in their work with brain-injured patients proposed that normal humans are capable of two modes of behaviour--abstract and concrete. These two modes are "capacity levels of the total personality". The abstract and concrete attitudes are not just mental sets or habits of the person, rather each one provides the basis for all performances relating to a specific plane of activity. Although normal humans are capable of both levels, Goldstein found that brain-injured patients were often deficient in the abstract attitude.

Concrete behaviour, according to Goldstein, consists of reacting to a stimulus in an automatic or direct manner. The behaviour is confined to the immediate apprehension of the given situation in its particular uniqueness. Most of our daily activities are run on the concrete basis, that is acquired performances which do not need conscious, volitional activity.

In the abstract attitude however, man can transcend the immediately given situation, the specific aspect or sense impression. He can detach himself from the given impression. The abstract behaviour is brought into play "whenever the situation cannot be mastered without the individual's detaching his ego from the situation".

Goldstein maintains that these are two separate levels of ability, and that there is a pronounced difference between the abstract and the concrete behaviour. The abstract level does not represent a gradual ascent from the concrete level. Abstract behaviour is not simply a combination of existing lower functions. There is a decisive difference between the abstract attitude of active shifting, active synthesis and the concrete attitude of passive global reactions to stimulus.

The abstract level is, according to Goldstein, characterized by the appearance of conscious will. This is an essential part of the new level. According to Goldstein, any other kind of seeming abstraction which does not involve conscious will is not abstraction at all. Goldstein then defines abstraction as the ability to carry out a conceptualization by conscious volition.

The normal individual functions on both the abstract and the concrete levels, and in his everyday activities they are intertwined in a fluid relationship, the person switching back and forth from one mode to another as needed. However, Goldstein found that brain injured patients have great difficulty in assuming the abstract attitude. The patient is reduced to a level of concreteness of situational thinking and acting, and can perform only those tasks which can be fulfilled in a concrete manner.

Some aspects of this impairment of the abstract attitude produced

by injury to the frontal lobes in humans are demonstrated by some individual examples given by Goldstein and his associates (1942 & 1944).
The patients varied in the nature and degree of their brain injuries and
as a result, not all were affected in exactly the same way. One patient
was asked to repeat the sentence: "The snow is black". The patient
stated that he could not say it, that it was not so. The examiner explained to him that such senseless phrases can be repeated even though
they are not true, and then asked the patient to repeat the sentence
again. This time the patient repeated the sentence, but mumbled immediately afterwards: "No, the snow is white". Another patient was
able to use eating utensils while eating, however when given these same
utensils outside the eating situation, was only able to produce a jumble
of senseless movements with them.

Another patient who had just been successful in reciting the days
of the week was then asked to recite the alphabet. He was unable to
shift to this task, and only after the examiner had commenced to call
out the alphabet was the patient able to follow in his recitation. One
patient complained, "I tried working out jigsaw puzzles but I was very
bad at them. I could see the bits but I could not see any relation between them. I could not get the general idea." Some patients were
able to read words correctly, but if the letters of the same word were
presented separately with a space between each letter, they were
unable to recognize the word.

Many patients could easily find their way when walking from the
ward into a room, but if asked to draw a map of their route, they could
not. One patient was able to recognize the testing room because, unlike the other rooms it has three windows in it. However, it was necessary for him to open a number of doors before he found the correct
room because he had no spatial orientation which would enable him to
locate the room with three windows. He was able to get to the shop or
dining room only by following the other patients. If he got separated
from them he became lost. He was able to recognize his own bedroom
only because he had tied a string to the bed post.

This same patient was once shown a picture of an animal and was
asked to identify it. When he could not decide whether it was a dog or
a horse, he addressed the picture directly saying "Are you a dog?"
When the picture did not reply, he became very angry. When given a
mirror he would look behind it for the person he saw in the mirror and
became excited when he did not find a person there.

Goldstein's work in this area has been most valuable in several
fields of psychology and physiology, including that of understanding and

helping persons with brain injuries. However, he was mainly concerned with differences between brain-injured individuals and normal
individuals. He did no comparative work between normal humans and
other animals, but he takes the position that while normal humans are
capable of both the abstract and the concrete modes, other animals
are only capable of the concrete attitude.

Although most valuable in other areas, Goldstein's extensive work
on brain-injured patients did not enable him to formulate a new definition of abstraction that would enable scientists to test other animals
for the possession of this mental ability. He stresses the importance
of conscious will in his definition of abstraction, but this is of no help
in testing other animals. Scientists have no way of knowing if a particular behaviour of an animal was carried out consciously or otherwise. Indeed, we do not even know whether they have what we call
conscious behaviour or not.

Gradually over the years, the difficulties in defining the term abstraction caused most researchers to abandon the term in favour of
"concept" or "concept formation" which has generally been given the
definition of "common response to dissimilar stimuli". There have
been several experimental studies designed to show that only man is
capable of abstraction or concept formation. However, because of different definitions of "abstraction", the interpretation of the results of
these studies is a matter of dispute.

Chimps as well as rats have been successfully conditioned to respond positively to triangles and negatively to circles. These reactions
although learned for certain triangles, proved to have also been learned
for other triangles as well. Munn (1955) asserted that the animals
"were responding to the different triangles in terms of some property
common to all, such as three-sidedness or triangularity". Thus, the
animals apparently formed the concept of triangularity. However,
Goldstein disputes this assertion. He says that the animals learned
"to react to common aspects on an entirely concrete basis" (1941).

Animals were shown, however, to have difficulties in learning to
discriminate the middle of three serial stimuli. Experiments with the
white rat and the dove, showed that it is much more difficult for the
animal to learn to distinguish the middle of three stimuli of varying
magnitude than to discriminate between two extreme stimuli (i. e. tri-
angle and a circle). Apparently the middle stimuli is "less likely to
stand in perceptual contrast with others (i. e. to be concretely impres-
sive) than the two extreme stimuli" (Goldstein, 1941, p. 25).

Although some of those studies may seem to suggest that animals

are limited to concrete responses, using the only accepted operational
definition of concept formation as "common response to dissimilar
stimuli", it is clear that animals are able to do just that. As yet, there
are no studies that are universally accepted as demonstrating a difference of kind between man and other animals as far as abstraction or
concept formation is concerned.

ART Every human culture has some form of art. And what is more,
no art from even the so called simple or primitive cultures can be
called immature. If we know enough about a culture to be able to understand and appreciate its art forms, we inevitably find a high level
of maturity and sophistication. Although not all societies have all of
the art forms we are accustomed to, and sometimes their art forms
are less well executed than ours--especially when their tools and techniques are inferior to ours--no art can be considered childlike or immature.
      It has long been widely believed that humans are the only animals
capable of producing these forms of behaviour. However, in the 1950's
there was a great deal of publicity about chimp "art". Experiments
were conducted by several scientists that produced ape and monkey
drawings and paintings. These works could best be described as scribbles and some of their paintings looked a great deal like "abstract"
art. Chimps are also known to have what might be called a crude sense
of music or perhaps a more accurate term would be a sense of rhythm.
Chimps can readily learn to beat a bass drum in time to simple jazz
music.
      Is this another case of one of our proposed unique abilities falling
under the hands of a chimpanzee? In order to answer this question,
we must first examine the nature and components of human art.

HUMAN ART Undoubtedly it is the aesthetic component which forms
the nucleus of any art form. An object or performance cannot properly
be called art unless there is included an element which is over and
above any possible utility or efficiency. This aesthetic element exists
solely for the enjoyment and satisfaction of either the artist or also
for those who may see it.
      Of course the more commonly accepted forms of art such as
paintings or sculptures have no functional utility--that is, you cannot
store water in them or hoe the garden with them. However, other objects, which we do not normally consider to be forms of art, such as
dresses or pottery can also become works of art if decorations or or-

namentations are included in their production. Few people normally
consider a rug to be an art object. However, the Navajo Indians have
developed the intricate designs and craftsmanship of their hand-made
rugs to the point where they are regarded as valuable art objects the
world over.

There is also an additional component of almost all, if not all
human art. An artist is usually trying to communicate something to
others who will be viewing his production. This communication may
be specific and conscious as when ideas are communicated in literature
and drama, or the communication may be more general and only semi-
conscious or even unconscious in art forms such as music or paintings.
It appears that some forms of art lend themselves better to the communication of ideas, whereas other forms are much better mediums
for the communication of feelings, emotions and moods. There are
many people who believe that all (human) art contains this element of
communication. Certainly the vast majority of our art can be shown
to be expressive of some type of communication rather than to be purely
aesthetic.

To the extent that the artist is able to succeed in communicating,
he must use the accepted conventions of the culture in which he is
working. These conventions are always present in every society, even
though they are often not explicitly stated as such. These conventions
are often very largely unconscious and we may have difficulty in recognizing them as cultural conventions until we study other societies in
which the conventions differ from our own.

Probably the conventions used by the literary artist or author are
most clearly recognized. He is bound by the language in use by the
society of which he is a member. He cannot hope to communicate anything to his audience unless he adheres rather closely to the linguistic
conventions or symbols in use in that society. Unlike most cultural
conventions, linguistic conventions, at least in modern societies, are
usually formalized and recorded in the form of dictionaries, rules of
grammar and so forth. Also, unlike other forms of artistic conventions, most people are easily aware that the reason they cannot understand the language of a person from a different culture is because of
the differences in learned conventions. Unfortunately, many people
tend to react to culturally foreign forms of art and music with the belief that these are in some way inferior or primitive or meaningless,
instead of by recognizing that these works are using different conventions, which the viewer must learn before he can appreciate them.

The more common use of unconscious conventions can be shown in the art forms of drawings and paintings. For example, in our culture from a very early age we learn to accept the convention of using a two-dimensional representation to communicate a three-dimensional space. Many people are not even aware that this is a learned convention. However the conventional nature of two-dimensional representations is shown by the fact that the people of some cultures such as in Oceania and Africa have no two-dimensional art and are at first not able to recognize or correctly interpret photographs and drawings.

When first shown pictures, these people often tended to react in one of two ways. Either they were completely unable to recognize the pictorial representation or else they reacted as if the picture was the real object itself. The first reaction was recorded by Robert Laws at the end of the last century in what is now Malawi, Africa.

> Take a picture in black and white and the natives cannot see it. You may tell the natives: "This is a picture of an ox and a dog," and the people will look at it and look at you and that look says that they consider you a liar. Perhaps you say again, "Yes, this is a picture of an ox and a dog". Well, perhaps they will tell you what they think this time. If there are a few boys about, you say: "This is really a picture of an ox and a dog. Look at the horn of the ox, and there is his tail!" And the boy will say: "Oh! yes and there is the dog's nose and eyes and ears!" Then the old people will look again and clap their hands and say, "Oh! yes, it is a dog!' When a man has seen a picture for the first time, his book education has begun. (Quoted in Beach, 1901, p. 468)

The second reaction is shown from another report when photographs were projected on a large screen for an African tribe.

> When all the people were quickly seated, the first picture flashed on the sheet was that of an elephant. The wildest excitement immediately prevailed, many of the people jumping up and shouting, fearing the beast must be alive, while those nearest to the sheet sprang up and fled. The chief himself crept stealthily forward and peeped behind the sheet to see if the animal had a body, and when he discovered that the animal's body was only the thickness of the sheet, a great roar broke the stillness of the night. (Lloyd, 1904)

Perhaps the best example of unconscious conventions in art can be indicated in music. The person of a Western cultural background finds music from another culture such as China to be meaningless, unpleasant and often is unable to understand how the Chinese are able to enjoy listening to it. Chinese not accustomed to Western music often make similar remarks about our music.

Chinese music, as most non-Western music, uses a different scale from our music. Any musical scale arbitrarily sets fixed intervals between notes. The Chinese music simply employs a different system of intervals between notes. It is not the pitch of the individual notes in Chinese music which Western listeners find inharmonious, but rather it is the interval between the notes that we are unaccustomed to.

In any culture when an individual artist or small group of artists first breaks with established conventions and substitutes new conventions, there is an inevitable period of misunderstanding and even rejection by the society. This pattern can be shown in our own culture for example in painting when the abstract school first began early in the century and also for music with the early development of jazz.

We see then that a very important component of human art is the element of communication. In this respect, art and music are similar to language and writing. Language and writing will be discussed more fully in the next chapter. However, it is already evident that art, music, language and writing are all forms of communication which rest their ability to communicate on the learned cultural conventions or symbols present in each culture. When trying to understand a different culture, it is not only necessary to learn their linguistic conventions in order to understand their speech, but it is also necessary to learn their artistic conventions in order to appreciate their art and music.

APE AND MONKEY ART   It is now possible to examine in more detail the "art" of apes and monkeys and see how it compares with human art. Morris (1962) has compiled the fullest account of the research done in this area with apes and monkeys. Most of the study has been conducted with chimps, but a few gorillas, orang-utans and capuchin monkeys have also been observed. As far as is known, apes and monkeys never make any type of drawings in the wild. In the laboratory, it was necessary to first demonstrate the use of a pencil or paints and paper to the animals. However once shown what to do, the animals carried on quite spontaneously and required no further assistance or guidance from the experimenters. In many cases this type of behaviour was carried out enthusiastically by the chimps. No reward was given to the apes for their picture making, apparently this behaviour is self-rewarding, that is the apes apparently enjoy doing it for its own sake. In some

Fan pattern drawing by female chimpanzee

cases, if the experimenter tried to take the drawing away from the
animal before it was "finished", the ape became extremely upset and
violent.

The scribbles of these animals develop and change over time as
they acquire experience. An individual ape usually starts with relatively simple lines, but with experience changes more and more to
mulitiple scribbles. There appears to be a distinct patterning and
sense of design employed in these scribbles. Each animal appears to
have an individual drawing style which seems to be fairly constant over
time. Any attempts by the experimenter to influence the kind of picture being produced were always unsuccessful.

From these results we can fairly confidently assume that the apes
are capable of at least some primitive degree of aesthetic enjoyment of
their artistic work. However, the most important fact in this area is
that none of these apes or monkeys were ever able to reach the imitative or representational stage of art. That is, none of these animals,
no matter how old or experienced, were ever able to draw even a crude
picture or outline of some object. A human child's scribbles quickly
become imitative, but a chimp's never do. In other words, they were
never able to formulate the conventions or symbols that were neces-

sary before art can become a form of communication.

There have been some studies done with chimps indicating that they are able to obtain some information from black and white photographs. The Gardner's chimp has been able to correctly identify photographs of many objects (see Chapter 2). However at least at first, chimps are apparently reacting to the photographs as if they were the real thing, rather than by realizing that they are only representations of the real thing. When Kohler (1927) showed his chimps photographs of other chimps they always reacted as if they were seeing another chimp. Most of the chimps tried over and over to see what was behind the print. The most intelligent of this group of chimps repeatedly raised his arm to the photograph in the typical chimp gesture of friendly greeting. How long chimps continue to regard photographs as the real object, rather than as a representation of the real object is not known. The evidence on chimp interpretation of two-dimensional representations is not clear at this time. However, it is clear that chimps have not been able to produce two-dimensional reproductions of their own.

As mentioned previously, imitative two-dimensional drawing is a learned convention which most, but not all human cultures use. The natives of some cultures in the South Pacific and Africa had not formulated the convention of using two-dimensional art. It must be made clear that there is no similarity between the lack of two-dimensional art in these people and the inability of the chimps to produce this type of art. Although the natives were at first unable to recognize two-dimensional art, when properly instructed, within a very short time these people were able not only to recognize work using this convention, but also to make use of the convention themselves. The chimps, on the other hand, although given ample opportunity to develop this convention in the laboratory were never able to do so. Also, as far as is known, they have never been able to produce this in the wild either.

CONCLUSION In conclusion, when we ask if humans are the only animals able to produce art, we must first define exactly what we mean by "art". If we mean only the aesthetic element when we speak of art, then it is apparent that apes are indeed capable of some simple form of art. However, if we include in our definition the component of representation or communication, it seems clear that here we have a genuine demarcation line between human ability and non-human ability. Humans are capable of producing representational art, other animals are not.

It is most unlikely that man's first attempts at art were representational. The earliest evidence we now have of human art are the "doo-dles" drawn with fingers on damp clay in a cave in Spain dated about 30,000 BC. Like today's chimp scribbles, these early human doodles were almost certainly not representational--in spite of some people's attempts to find pictures in them.

As far as the beginning of human representational art is concerned, it is of great surprise and mystery to anthropologists that the present evidence indicates a sudden outburst of mature and sophisticated rep-resentational art around 20,000 years ago in the upper Paleolithic pe-riod. The excellence of these drawings in the caves of France and Spain is remarkable because of the lack of accompanying primitive or immature art. One might expect to find large quantities of "beginner's" art in this area also, but one does not.

After some reflection on the previous discussion of the nature of human art, the sudden appearance of high quality art may seem some-what less mysterious. It may well be that once mankind formulated the necessary conventions for producing representational art, imple-mentation of this activity reached a mature level in a very short time. As the evidence from the ape scribbles and the lack of two-dimensional art in some cultures seems to indicate, apparently the truly difficult part in developing art was the formation of the conventions or symbols that are necessary before the production of imitative art can begin.

Humans must first have acquired the ability to make such formu-lations through evolution (apparently chimps still do not have this ability), and then some time later to first formulate the actual conven-tions which enabled representational two-dimensional art to begin (which some cultures still have not done). Apparently once these con-ventions were formulated, the implementation of art was relatively easy and reached a high level of sophistication in a short time.

CULTURE With the growth of the new science of anthropology, it has become popular to remark that only mankind has culture. However, the serious student of anthropology realizes that anthropologists can not even agree upon a definition of culture. Although most anthropolo-gists accept that humans are the only animal with culture, a few are not so certain.

It is not necessary to burden the reader with all the competing definitions of culture put forth by the various schools in anthropology. It is sufficient to list the characteristics which all anthropologists would agree are necessary for a minimum definition of culture. Cul-ture is dependent upon the ability of one animal to acquire or learn

from another animal information previously learned by the first animal.
This ability to pass on knowledge and skills from one individual to another individual and therefore from one society to another, or from one
generation to the next must be at the core of the meaning of culture.

The tremendous advantage given to a species that has culture is
that it eliminates the necessity of each animal learning every problem
by "trial and error". Once one animal has learned a skill or useful
behaviour, this information can be passed on to other members of the
group. It is no longer necessary for each individual to learn everything first hand or by trial and error. Learning is acquired by the
group rather than by only the individual.

Also it is necessary to emphasize that cultural behaviour must be
learned, it cannot be genetically transmitted. Culture must be non-
biological. Behaviour in other animals which might at first appear to
be similar to our own cultural activity, can often be shown to be rather
the result of biology. For example, greylag geese pair off permanently
for mating. Since this is similar to the practice in many human cultures (ideally), it might be tempting to call this behaviour cultural.
However, it can be shown that this behaviour in the geese is not learned,
that it is controlled by genetics. One consequence of this is that all
flocks of geese are identical in this respect. Each flock has the same
behaviour in this activity as do all the other groups. Since the behaviour is controlled biologically, there is no room for group variation.
In contrast, the behaviour of humans in different societies in regard to
mating differs markedly. Indeed, Homo sapiens behaviour shows a
truly remarkable variation in many activities from one society tp the
next. For any behaviour to be considered cultural, it must be shown
that this behaviour differs from one group to the next. A shorthand
definition of culture then, might be stated as pooled learning or non-
physical inheritance which is different for each society.

Obviously, humans are cultural animals *par excellence*. We communicate knowledge to each other on every conceivable subject and
skill. Evidence indicates that our culture traditions started very slowly,
with new discoveries or inventions sometimes taking hundreds of years
to spread across a continent. But slowly, this process of acquiring
information from other humans speeded up. Also the rate of learning
has increased dramatically. Humans have at least to some extent,
learned how to learn--we have slowly acquired a "learning set" about
how to learn at least certain types of knowledge. So that today, not only
is our rate of discovery much faster, but also once a discovery is made,
this knowledge can be transmitted around the world in as little as a few

hours.

It is evident that nearly all our cultural activity is possible because of our symbolic language. By word of mouth or writing or even by pictures, humans pass learning on to other humans. If it were not for the use of symbolic language, much of this transfer would not be possible. As will be discussed in the next chapter, no other animal is able to communicate to another learned information or skills by the use of symbolic language.

However, it must be pointed out, saying that other animals are not able to pass information on by the use of symbolic language, and saying that no other animal has the ability to transfer learning are not quite the same thing. It is possible that other animals could transfer learning to others *without the use of language.* And in fact, this is just what some animals are able to do on a very limited scale.

Without language, the capacity to transmit previous learning is dependent upon imitative learning. That is, seeing another individual performing a new action and then copying that action. Humans of course can do this, but apes, monkeys and possibly cats are also capable of imitative learning.

When Yerkes (1943) first set up his laboratory in Florida, the pioneer chimpanzees were shown how to work the drinking fountains. Through the years, it was never necessary to instruct the new generations of chimps on the operation of the fountains. Each new chimp learned how to perform this task from watching another chimp. Because of this and other examples, Yerkes did not consider chimps incapable of culture. Chimps are the best among non-human animals at this type of imitative learning. There are numerous examples showing how readily they do this; for instance, they can learn to operate a simple set of latches merely by seeing another do this act. Many other examples of imitation can be found in Kohler (1927).

Monkeys are also able to learn from one another by imitation. A particularly interesting example was found by Kawamura (1959) among the wild Japanese Macaques. The practice of washing sweet-potatoes with water before eating them was started by a three year old female. The practice was first transmitted to her playmates and then to the mother. New habits such as this usually start among the infants and then spread to the older members by what Kawamura calls "acculturation".

There is definitely variation among the wild Japanese macaque troops as to the types of food consumed. One troop knows how to remove a certain type of root from the earth, while another troop en-

tirely lacks this knowledge. Another troop frequently invades rice-
paddies, yet another troop never damages them in spite of their living
many years near rice paddies.

It seems clear then that at least apes and monkeys are capable of
transferring information by imitation. Does this mean they have a
primitive form of culture? It is a matter of opinion. In order to give
man the exclusive claim to culture, it is necessary to include language
somewhere in the definition. This author is of the opinion that language
should be considered in its own right. The ability to use symbolic language and the possession of culture are not the same thing. Each
should be considered separately. True, most of our culture depends on
the use of symbolic language, but not all of it. Even if we are the only
animal able to use symbolic language, that does not automatically mean
we are the only animal with culture.

The amount and type of learning that can be transferred by imitation is very limited and simple. Ape and monkey "culture" is so primitive that it is tempting to dismiss it altogether. However, ape tool-
making compared to our tool-making ability is almost ludicrously simple. Sticks and home-made sponges are hardly comparable to com-
puters and lasers. Nevertheless, sticks and home-made sponges *are*
tools. And however simple, these tools mean that we can no longer
call ourselves the only tool-making animal.

Likewise, even though ape and monkey transfer of learning by imitation is almost absurdly primitive compared to our cultural activities,
we cannot deny that some transfer of knowledge is taking place. And
painful as it may be for us to admit it, we can no longer claim to be the
only animal capable of cultural activity.

HUMOUR In the middle ages it was often said that man is a rational
animal who laughs. Perhaps outright laughter is unique to humans.
Kohler (1927) reported that he had never observed apes laugh or weep.
However they sometimes produce a rhythmic gasping or grunting, somewhat resembling our laughter when they are being tickled. Also, when
doing something that gives them particular pleasure, they form an expression that resembles our "smile". Although Kohler believed that
with experience, chimps are able to correctly interpret human facial
expressions, he states that they were never able to understand merry
human laughter.

It seems that chimps do appear to have a crude sense of humour.
Humour is usually dependent upon recognizing the incongruities or
peculiarities in a given situation. In order to recognize these incon-

gruities an animal must in some sense be detached or removed in his thinking from the immediate situation.

Chimps in zoos sometimes lure visitors near their cages and then shower them with saliva or handfuls of mud. The chimps then run around their cages in noisy glee. Kohler tells how chimps would sometimes tease hens by holding slices of bread through the wire fence, waiting until a hen would approach the bread and then pulling the bread back at the last moment. At one meal, this "joke" was repeated around fifty times.

These examples are enough to show once again that chimps have dethroned another supposed unique human ability. Although most interesting, there has been little if any comparative experimental work on humour in various apes or monkeys. Humour because of its fluid and fleeting nature has always been difficult to study scientifically even in humans.

SELF CONCEPT Psychologists have long considered humans the only animal able to form a "self concept". This term is usually defined in such a way however, as to make it impossible to test other animals for the possession of this ability.

Recently it has been suggested by Gallup (1971) that the ability to recognize oneself in a mirror might be one aspect of the ability to form a self concept. Even though most psychologists would consider this to be a very limited definition of self concept at best, it has the advantage of allowing us to examine other animals for the possession of this ability. This definition is so far removed from the usual one however, that perhaps it would be best if the term "self concept" were not even used.

Human babies usually show that they are able to recognize themselves in mirrors when they are about 10 months old. Retarded persons may never acquire this ability. It will be recalled that one of Goldstein's brain injured patients was unable to recognize himself in a mirror and tried to look behind it to find the person he saw.

Most animals when first confronted with a mirror will at first react as if another member of their species stood before them. They soon lose interest in the mirror however and never show signs that they recognize themselves.

Chimps (but not monkeys) react differently however. At first they react as other animals, that is responding as if another chimp were in front of them. Kohler's chimps would often grasp behind the mirror apparently trying to catch the chimp they saw before them. Gallup found however, that on the third day after first being exposed to a mir-

ror, they began to show they were able to recognize themselves by inspecting parts of their bodies they could not see otherwise and by
making grotesque faces at themselves. Kohler found that unlike other
animals, a chimp's interest in mirror images did not decrease with
time.

In some ways this ability to recognize oneself in a mirror seems
more similar to recognizing two-dimensional photographs than it does
to "self concept". As will be remembered, Kohler's chimps also reacted to black and white photographs of chimps as if they were observing
another animal. A mirror image is in colour and makes identical moves
simultaneously with the observing chimp, whereas of course a photograph is always stationary. A chimp is able to recognize itself in a mirror after two days, but Kohler did not mention if his chimps were ever
able to react to photographs except as if they were other chimps.

At any rate, it seems that only chimps and perhaps other apes
among non-human animals can recognize themselves in a mirror. The
relationship (if any) of this ability to "self concept" is difficult to assess.

FORESIGHT From time to time it has been proposed that man is the
only animal that has foresight, the ability to anticipate what will be
likely to happen in the future. Indeed, a large proportion of man's
energies are spent on activities which are preparations for the future.

No modern psychologists would defend, however, that man is
unique in this ability. McDougall (1928, 1931) found that all animals
anticipate the future at least to some extent. Thorpe (1963) concludes
that an element of expectancy or foresight "is as fundamental to organisms as is perception of space" (p. 115). A rat in a Skinner box
pressing a lever for food certainly acts as if it knows what its actions
will lead to.

Although no psychologist or ethologist would claim that man is
unique in his ability to anticipate the future, many would agree that
mankind's foresight is certainly far more extensive than that of any
other animal. As yet, no one has succeeded in discovering if man's
foresight is different in kind from that of the other animals, and, if
this is so, exactly where and why the difference exists. Human foresight will be discussed in a great deal more detail in Chapter 5.

There is, however, one remaining proposed difference which has
yet to be discussed. This proposed difference is of such importance
that it is given a chapter all to itself.

Chapter 2

Language

It is not known when men first proposed that humans are the only animal capable of using language. Since language is the most obvious and noticeable of the proposed differences between man and the other animals, it is quite likely that this difference was first advanced very early in our past--possibly not long after our ancestors first acquired speech. Certainly some of the "primitive" cultures of today such as the Australian aborigines believe that language is one of the basic differences separating humans from all the other animals.

Not only is language most likely the earliest proposed difference, but also it is today the last stronghold of the scientists who believe that man is different in kind from the rest of the animal kingdom. The only remaining proposed difference today vigorously defended by reputable scientists is that of symbolic language. A large number of writers have proposed that it is language which separates humans from the rest of the animal kingdom. Under the careful scrutiny of scientific observation and experimentation the other proposed differences have either been disproved or simply abandoned for lack of evidence.

Broadly speaking, proposals that our language differs in kind from that of other animals have fallen into two classes. First, many people such as Cassirer (1944), White (1949) and Langer (1972) have suggested that it is the symbolic nature of our language that distinguishes us from other animals. The words we use are *symbolic* of what we are talking about. The second line of attack has been taken mainly by Chomsky (1965 & 1968) who argues that it is the nature of our sentences which is important. The way we organize our words or symbols into sentences is the crucial difference between human and animal language.

This is not to say that everyone agrees that language does indeed mark man apart. As usual, there are those who believe that although human language is certainly far more complicated and extensive than any form of communication used by other animals, it is not different in kind from animal communications. Recent experiments in teaching chimps various human languages have bolstered the stock of these

people in the "degree" camp. These experiments will be discussed in
more detail later in this chapter.

Certainly there is agreement that other animals can and do communicate with each other (i. e. transfer information, emotion or mood
from one animal to another) on at least some level. What is in dispute
is the nature of animal communication and how it compares with human
language. Is there something about human language that sets it apart
from any other type of animal communication? Or is our language simply a more complicated version of the same thing other animals do to
communicate?

The proposal that it is the *symbolic* nature of our language which
marks us as unique will be the main proposal discussed here. What
exactly do we mean by "symbolic language"? Why is only human language said to be symbolic? Instead of trying to give a full definition of
symbolic language at this time, perhaps it would be best to first point
out two important aspects of our language that will enable us to understand the difference between human language and most forms of communication used by other animals.

First of all, it is important to realize that human language is
learned, that is it is not genetically determined. Human language must
be learned or acquired by the young. A human child that was not exposed to a language would not have the use of language. Apparently
there is a critical age for learning language and if a child is not exposed to a language during this period, it is much more difficult for
him to learn a language later in life.

A second important aspect of almost all human language is that it
is arbitrary. That is, the sound or signal used as a word for something else is in no way similar to or an imitation of the original object.
For example, the word "horse" is in no way suggestive or imitative of
a real horse. Say the word "horse" to a Chinese, and he will have no
idea what you are talking about. Human language is arbitrary in that it
arbitrarily associates a word with an object or event without trying to
link up those objects with those words that are in some way similar.

Obviously there are a few words in each language which are not
arbitrary. Onomatopoeic words such as "hiss" or "boom" or "moo"
are examples of some words that are suggestive or imitative of the
original event. Say "moo" to a Chinese and he will likely have some
idea of what you are trying to communicate (he will also probably think
you mad). With the exception of these few imitative words, nearly all
human language is arbitrary.

Given these two important features of human language, that it is

learned and that it is arbitrary, we can already begin to select certain
types of animal communication that are universally agreed to be different in kind from human language. These two types of animal communication can be called genetic or unlearned signals and gestures.

GENETIC, UNLEARNED SIGNALS

The fact that some animals are able to communicate with each
other by using elaborate signals is now commonly accepted. Often however, these communications can be shown to be genetically determined
for the species. That is, they are not learned or acquired by experience by the young animal as is the case with human language. Rather
the "language" is a hereditary faculty present in adults even if they
were kept isolated from adult members of their species when young.

Probably the best known example of this type of animal communication was discovered by Karl von Frisch (1962) in the honeybee. The
honeybee is able to communicate certain specific types of information
to other members of the hive by means of a "tail-wagging dance".
When a food foraging bee finds a new source of food, she returns to the
hive and is able to communicate information of the distance and even
the direction of the food to other members of the hive. Generally
speaking, the richer the source of food, the longer and more vigorous
the dance.

For food found only a short distance from the hive (less than 100
metres) a "round dance" is used by the bees. This dance does not
communicate the direction of the food or the distance, except that it is
near the hive. For food found farther away, the bees resort to the now
famous "tail wagging dance".

The finder bee starts this dance by running a short distance in a
straight line and wagging her abdomen at the same time. Then she
turns and returns in a semicircle to the starting point. She then repeats this but returns using a semicircle of the opposite side. This
dance is then repeated many times. The bee is able to communicate
the distance of the source by varying the tempo of the dance. The
farther away the find, the slower the dance. For example, when the
find is only 100 metres away the bee makes 10 complete runs through
the pattern in 15 seconds, but when the find is 1,500 metres away, the
number of runs drops to 4.5.

Communicating the direction is more complicated. In essence
however, the indicating is done by using the sun as a reference point.
During the straight part of the dance, the bee indicates the relative
angle of the find in relation to the sun. The other bees are then able to

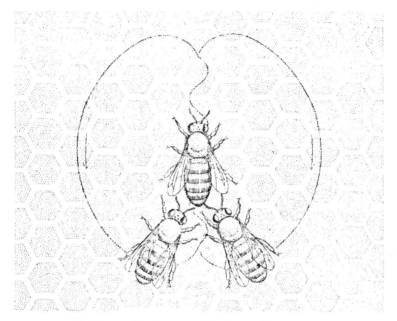

Tail-wagging dance of the honeybee

fly off at that same angle towards the food.

The "tail-wagging dance" of the bees is by far the most complicated and information laden "language" used by any other animal except man. But for our purposes, the important point about the dance is that it is not learned. When young bees were taken from the honeycomb and raised out of contact with the older bees, they were immediately able to communicate with the others after being returned to the hive. The bee dance is largely genetically determined.

More evidence that this language is not learned can be provided by putting different varieties of the bees together. The Italian variety and the Austrian variety use slightly different systems for communicating the information during the dance. When von Frisch put the two varieties together, they were able to work together, but confusion arose about the dance communications. These two varieties of bees were never able to learn the language of the other.

Other animals also use this type of genetic communication system. Birds are able to relay precise information by means of "call notes". Thorpe (1961) has shown that birds make two basic types of sounds. The best known type of bird sound, "songs", are sometimes learned or

partly learned depending upon the species. These "songs" are not used
to convey information however. The second type of sound used by birds
is that of "call notes". These call notes are used to sound a warning
of the presence of a dangerous bird such as a hawk or owl. The call
notes can be shown to be genetically fixed similar to the dance of the
honeybee.

Varvet monkeys have three different warning calls. Seyfarth,
Cheney & Marer (1980) found that one call is given to warn of a nearby
dangerous running animal (such as a leopard), another call for airborne
danger (such as an eagle) and yet another for nearby ground danger
(such as a python). Monkeys hearing these warning calls take appro-
priately different defensive reactions to each call. There is a certain
amount of learning involved here as young monkeys will emit appropriate warning calls for dangerous as well as non-dangerous animals, but
adults learn to discriminate and make these calls only when strange or
dangerous animals are seen. However, the warning calls themselves
appear to be unlearned.

Birds and other animals also use other types of genetic signals in
additon to calls--postures, movements and displays of colour are used
to elicit appropriate responses from other members of the same species. Again these signals have proved to be largely unlearned or highly
"environment resistant". Many species from fish to apes use this type
of genetic signal for communication. The number of signals used by
each species is severely limited by our standards, however, less than
40 even in the monkeys and apes. Moynihan of the Smithsonian Institution has indicated that the largest number of these genetic signals
used by a species is 37 in the rhesus monkey.

Genetic signals may or may not have arbitrary meanings. Some
genetic signals are similar or imitative in form to the event being communicated (such as when animals growl and bare their teeth at each
other). The meanings of other genetic signals are arbitrary, such as
the call notes of birds. There is nothing about the sound of the call
note that resembles the sound (or any other aspect) of the dangerous
bird in the vicinity. The connection between the dangerous bird and the
call note is purely arbitrary. Genetic signals may or may not have
arbitrary meanings, but they are always unlearned.

The limitations of this type of communication are clearly evident--
first of all the small number of signals used by each species indicates
the severely limited type of communication that can be accomplished.
With such a small number of "words", not much can be said. The second disadvantage of genetic signals is probably even more impor-

tant--since these signals are genetically fixed, it is not possible to add new or change signals rapidly when environmental conditions change for a given species. Any changes that occur in genetic signals must occur through the painfully slow process of biological evolution.

Human languages by contrast, are marvellously rapidly adaptable and flexible. Not only does each human language have thousands of different words which can be used to communicate almost any conceivable message, but it can also be shown that each culture's language has been and is still being moulded to fit the needs of that particular culture's environment. In the English language for example, we have only one word for "snow", but snow is much more important to the Eskimo people and they have over twenty precise words to describe the different conditions of snow. If a sudden change occurred in a certain society's environment and the present language of that society did not already have the necessary words to describe this new situation, the words could be added, or slightly new meanings to old words could be employed, within a few days if necessary, to deal with the new situation.

GESTURES OR IMITATIVE SIGNALS

The second type of animal communication that is generally agreed to be different in kind from our language consists of gestures. A gesture is usually defined as a movement by the hands or body that is suggestive or imitative of what is trying to be communicated. Several higher mammals use gestures. Dogs often use mimic actions or gestures. When a dog wants us to come play with him, he will often leap and run ahead and then look back--imitating what he wishes us to do. Most people have experienced the gesture of a dog that wishes to be petted--poking his cold nose under our hand, trying to push our hand up on his head. Again this signal is an imitation or copy of what the dog wishes us to do.

Chimps do a great deal of gesturing. When one chimp wishes another chimp to come near, they use a beckoning gesture with their hand that closely resembles our own. They also have a begging gesture-- holding one of their palms up towards another animal or person. Chimps enjoy being tickled a great deal; when they wish to be tickled, they will often take hold of a person's hands and place them against their own ribs.

In all these cases, the gestures were imitative of the desired action. In other words, gestures are not arbitrary signals. Gesture signals of necessity must be closely similar in form to the original object or action. The meanings of gestures cannot arbitrarily be assigned. So we see that gestures violate the second important feature of human language--that of arbitrary meanings. Nearly all human language is arbitrary, but no gesture is arbitrary.

The term "gesture" is usually restricted to mean body motions, however in this book a slightly modified definition of gesture will be used. A gesture will be used to describe *any* acquired signal--vocal or body motion--that is not arbitrary. For our purposes, the important aspect of a gesture is that it must closely resemble the original object or event in some way. Thus, using this definition, it is also possible to have vocal or word gestures. Any word that does not have an arbitrary meaning such as "hiss" or "boom" will be defined as a vocal gesture. Also this means that a few body motions to which humans have assigned arbitrary meanings such as some hand signals used by deaf people will not be defined as gestures.

The limitations of using gestures, like genetic signals, are great. Because a gesture must closely resemble the original event, the amount and type of information it is possible to convey is very limited. In this respect gestures are similar to imitative learning discussed in the last chapter. Only simple, concrete events can be successfully imitated.

An interesting experiment with chimps by Crawford (1937) demonstrates how limited gesturing can be. Several chimps were in a cage,

a box containing fruit was brought up near the cage, but still out of
arms' reach. The box was rigged up in such a way with two ropes so
that it was possible for the chimps to pull the box close to the cage if,
and only if, two chimps pulled separately but at the same time on the
two different ropes. A few of the chimps were taught how the system
worked. Once they realized the necessity of cooperation, they proved
clever at watching one another and timing their pulls together. From
then on, once the box was set up and shown to one ape, he would get a
partner (often with a beckoning gesture) who was familar with the system, and together they would quickly pull in the prize.

However the most interesting part of this experiment was when the
first chimp tried to get help from another chimp that did not "know the
ropes" (no pun intended). No amount of frantic gesturing and imitation
could convey to the new chimp what action was necessary. The type of
information required to be communicated was beyond the range of
gesturing.

Unlike genetic signals however, gestures are flexible and can be
modified to fit the situation. Many animals use genetic signals, but
apparently only higher mammals are able to use gestures.

Genetic signals may or may not have arbitrary meanings, but they
are always unlearned. Gestures are acquired or learned, but they are
never arbitrary. In some cases it may initially be difficult to tell
whether a particular signal is a genetic signal or a gesture, but by experimentally isolating some young animals of the species, it should
always be possible to classify the signal as one or the other.

Except for the smile or crying in infants, humans use few if any
genetic signals. Together with the chimps we use a rich variety of
gestures however. We excel in yet another type of communication
however--that of symbolic language. If we do have the use of a unique
form of communication, it must lie with our use of symbolic language.
It is this form that we will now consider.

SYMBOLIC LANGUAGE

Humans communicate mostly through spoken and written language.
And it is our language that has long been believed to make our mental
activity different in kind from all the rest of the animals. To distinguish our language from other animals' communications, we often
designate it as "symbolic language".

Many philosophers and scientists still believe that humans differ
qualitatively from other animals because of our use of symbolic language. Several modern philosophers and anthropologists have ex-

pounded and defended the view that man has the unique possession of symbolic language (see Cassirer 1944, White 1949 and Langer 1972). I will discuss in some detail Leslie White's ideas here, as in my opinion, his theory is the most precise, straightforward and potentially scientific of the three authors mentioned.

White defines a symbol as "a thing the value or meaning of which is bestowed upon it by those who use it" (1949, p. 25). The meaning of a symbol according to White is never derived from or determined by properties intrinsic in its physical form. Thus, in my terminology a symbol must have an arbitrary meaning. A symbol can never be imitative or suggestive of what it refers to. A symbol can never be a gesture.

Almost all our words are symbolic. This is why foreign languages are almost totally incomprehensible to us. Listening to someone speaking in a foreign language such as Spanish, we are able to grasp precious little if any of the meaning of these strange words. There is nothing in the sound of the Spanish word "gato" for example, to suggest to us what it might mean. The word is in no sense imitative of what it refers to. We are at a loss even to guess what it might mean.

Often we lose sight of the arbitrary nature of nearly all our words. We are so accustomed to hearing our spoken word or seeing our written word for "cat" used to refer to a real cat, that we automatically think of a real cat whenever we hear or see the English symbol for this small animal. Our English symbol "cat" has been used so often in conjunction with the real animal that we sometimes seem to assume there must be a necessary connection between our symbol and the animal. But of course just the oppisite is true. There is absolutely nothing in the sight or sound of our three letter symbol that is in any way or form similar to or imitates or expresses the essence of the sight, smell sound, taste or touch of our feline friends. There is absolutely no natural or necessary connection between the two. Rather, it was only our Anglo-Saxon ancestors who decided that there should be such a connection. We could just as easily be using any other symbol such as "mijo" or "retta" or even "gato". We arbitrarily assign a connection between a symbol and an event or object.

As mentioned previously, there are a few words in each language that do not have arbitrary meanings. Properly speaking, these words are not then symbols but rather only vocal gestures. The English word "cat" is a symbol, the word "meow" is a gesture.

To distinquish between the way men use language and other animals use our language, White makes a distinction between a symbol and a "sign". White says humans can use both symbols and signs, but other

animals can use only signs. In some ways White's term "sign" is similar to my term "gesture", but not exactly.* His definition of sign
leads to some rather questionable and confusing statements however.
He says that in humans, symbols in some way can "become signs
through experience, as a consequence of repeated usage" (1973, p. 4).
He does not explain exactly how or when this happens or how it might
be possible to test and see if a word is being used as a sign or as a
symbol.

He also states that to other animals, a word can only be a sign.
"Dogs and apes understand words as signs, not as symbols" (1973,p 5).
Thus he believes that when a human learns a new word, he learns it as
a symbol, but when an animal learns this same word, the animal learns
it as a sign.

These statements are not possible to defend scientifically. Unless
it is possible empirically to distinguish between when a word is being
used as a sign or as a symbol by humans, or unless it is possible to
test and determine that animals are actually learning a word as a sign
rather than as a symbol, then we must call these statements scientifically meaningless. Because of the questionable statements reached as
a result of the definition of the term "sign", I have decided to abandon
the term in this book.

White's basic arguments however are still sound. By and large,
it is only his definition of "sign" that leads him into trouble. To avoid
confusion, I have decided to substitute the term "gesture" instead of
trying to re-define "sign". I hope this will lead to a clearer and more
defensible statement about symbolic language.

The term *symbol* will be defined here as any learned or acquired
arbitrary signal. A *gesture* on the other hand is an acquired imitative
signal. Using these definitions, a word such as "cat" is always a symbol no matter who or what is using it and likewise a word such as
"boom" is always a gesture. There is no mysterious shifting of a word
from symbol to sign depending upon who is using it.

A gesture should be understandable without any previous learning
or acquaintance with the signal, but a symbol never is. We might understand if a foreigner was gesturing to us that he wanted something to
eat, but we would not understand if he symboled to us in his foreign
language. A gesture by definition is similar enough to the original
event to allow the meaning to be grasped by what some psychologists

---

* White defines "sign" as a "physical thing or event whose function is
to indicate some other thing or event (1949, p. 27).

might call "generalization".

However, in order to understand a given symbol, previous learning must always occur. How is this accomplished? The learning of a symbol occurs by repeatedly pairing the symbol with the event. A parent points to a dog and at the same time repeats the word "doggie". After hearing the word "doggie" paired with the sight of a certain type of animal repeatedly, the child soon learns the connection. An adult also learns new symbols in a somewhat similar, but more complicated way. By looking up new words in a dictionary or by reading or hearing new words several times in a certain context, adults are able to learn the meanings of new symbols.

Humans are able to learn thousands of symbols. But what about other animals? White claims that humans are the only animals able to learn symbols, and that other animals can only learn signs. However, using my definitions, most if not all animals are able to learn at least a few symbols. Through repeated pairings, other animals are able to learn our symbols in a way similar to the way we learn symbols. Say "heel" often enough to a dog when he is standing at your feet and he will soon learn the connection. True, the number of symbols other animals are able to learn is small compared to our own--a few hundred at the most--but nevertheless, they are able to learn some of them.

So then, is the myth of our unique symbolic language gone in a puff of smoke? No, not yet. There is still one very important question remaining. And that question is--how did we acquire our languages in the first place? There are literally thousands of human languages in existence in the world today, not to mention all the "dead" languages-- languages that are no longer being used. We have records of only a few of these dead languages, most of them have vanished without a trace. How did they all start? The answer is we created them. We created them out of thin air.

Symbols are signals which have a meaning arbitrarily assigned to them. Not too many hundreds of years ago, not a single person on earth could have told you what the word "cat" meant. The English symbol for this animal did not yet exist. As a matter of fact, none of the English symbols existed at one time. The English language did not yet exist.

True, many other languages did exist before English, and English has borrowed heavily from several of these older languages. But trace any modern language back far enough and sooner or later it disappears.

The point is, humans create symbols. We easily create new symbols whenever it is necessary or pleases us. We are able arbitrarily

to assign a connection or relationship between a signal and an event or
object where once there was none. All new symbolic words arise in
this way. If you had walked up to anyone a few years ago and asked
them what the word "hippy" or "astronaut" meant, they would not have
known. These symbols did not yet exist.

Today we often lose sight of our ability to produce or create new
symbols. The English language already has so many thousands of symbols that it is nearly always possible to say anything we wish by using
only these pre-existing symbols. English has so many words already
that it is seldom necessary to create new symbols. Generally speaking
it is only in the fields of slang and new scientific discoveries that new
symbols are now usually produced. "Laser", "television" and "tran-
sistor" in the new discoveries department and "turned on", "right on"
and "far out" in the slang section are some examples of recently created symbols.

The net result of all this is that the average person seldom needs
to "symbolize", that is to create new symbols. Someone else has already created nearly all his needed symbols for him. But this is not
to say that symbolizing is difficult or requires any special education or
skill. When desired, any person can symbolize quite easily. Probably
the most common example of symbolizing in the lives of most people
occurs in the act of naming a new child or a new household pet. "What
shall we call him?" is a common question when a new member of the
family is expected. After some thought, a symbol is selected and
"given" or assigned to the new child or animal. Usually the name selected is completely arbitrary. Except for the question of the child's
sex, parents do not wait and inspect the child and then try to select a
name that in some way "fits" the child. Rather, the name is often selected before the child is even born. The act of naming is an act of
symbolizing. To name something is to say that there is now a connection between a certain signal and an object or event, where formerly
there was no connection.

Children are able to symbolize apparently just as well as adults.
Ask a child what a new pet should be named and he will have no trouble
giving you suggestions. Actually it may be that children have more
opportunity to symbolize than adults. Children often symbolize in their
play, creating secret code words and so forth. Recently an interesting
case has been reported (see Scobie, 1979) of a pair of twin girls in Cal-
ifornia who have created their own language! The twins were somewhat
isolated from about the age of 1 year, and apparently as a consequence,
created their own private language which scientists are now trying to

learn.
    Humans can easily create new symbols. We have done so millions
of times in the past, and we can do so as often as we like in the future.
All our languages were created by humans symbolizing and without this
ability, our languages would not now exist. Now we come to the most
important question in this chapter--can other animals symbolize? The
answer is no, they cannot. Other animals can learn a few of our symbols, but they have never been able to produce their own. Not once,
either in a laboratory or in the wild, have we witnessed an animal creating a single symbol, not one. Any human child can symbolize easily,
but other animals cannot.
    We have already seen that both humans and other animals are able
not only to understand gestures of others, but also to produce new gestures of their own. Other animals are able to produce their own gestures, but they are never able to produce their own symbols. Other
animals are able to learn some of our symbols, but they are never able
to create their own.
    Learning a symbol consists of learning a pre-existing relationship
between two dissimilar events. If these two events are paired together
often enough, almost any animal is capable of learning the connection.
Creating a new symbol or symbolizing however, is a completely different process. Symbolizing involves deciding to form or assign a relationship between any two dissimilar events that *have not been paired
together previously.* Learning a symbol consists of simply learning a
connection between paired events, symbolizing is assigning a connection
betwen non-paired events.
    Here then, we have a genuine difference of kind. We are able to
create symbols, but other animals have never been able to do so. If
any animal at any time had been able to create a single symbol, then
this would be a difference only of degree, but as far as we know, they
have not been able to. We are alone in our ability to symbolize.
We are different in kind.

RECENT EXPERIMENTS WITH CHIMPS
    Primarily because of several recent experiments conducted in the
United States attempting to teach chimps various languages, it has been
widely speculated that perhaps man's last claim to uniqueness--our
symbolic languages--was also about to crumble. These experiments,
conducted mostly by David Premack (1970, 1977), R. Allen & Beatrice
Gardner (1969), and Roger Fouts (1972), are most interesting and in-
structive as to the nature and limits of our nearest relative's ability to

use language. These experiments have shown once again that chimps
have more intelligence than we formerly gave them credit for.
    Until quite recently, it was thought that the main reason chimps
had so much trouble learning our words was because they did not have
the necessary intelligence. With a great deal of devoted effort, Hayes
& Hayes (1951) were able to teach a home-raised chimp to say only
three of our words (papa, mama and cup). Since chimps are known to
be able to produce a considerable variety of sounds, it was assumed
that their difficulty in learning our words was the result of a lack of
intellectual ability.
    However, more recent experiments have shown this assumption to
be false. Apparently the nature of chimp vocalizations is quite different from that of humans. Chimpanzees do make many different sounds,
but generally they occur only in situations of high excitement. When
undisturbed, chimps usually remain silent. When chimps do make
sounds, they tend to be specific to the exciting situations.
    This problem of chimp vocalizations has been overcome however
by the recent experiments in teaching chimps to use hand manipula-
tions and signals. Premack, by using small plastic bits varying in
shape, size, colour and texture to represent words, was able to teach
his chimp, Sarah, the meanings of over 120 words. The Gardners and
Roger Fouts have been able to teach the chimp, Washoe about 130 hand
signals. They taught Washoe some of the same hand signals that are
used by deaf persons in America, called the American Sign Language
(ASL). Fouts is also now teaching other chimps these same hand signals.
    The plastic bits used by Sarah all had symbolic meanings. Neither
the shape nor the colour of these plastic words were imitative of the
original object. For example, a banana was represented by a small
red square.
    The hand signals used by Washoe are mostly symbolic. The ASL
uses different hand signals that correspond to particular words or concepts. A large proportion of the ASL hand signals show some slight
similarity to the original object. For example, the ASL signal for
"cat" is produced by bringing the thumb and index finger together near
the corner of the mouth and moving them outward (to represent the
cat's whiskers). However, most of these hand signals are nevertheless
symbolic. Remember that a gesture must be similar enough to the
original object to allow communication without any previous training.
Although the ASL signal for "cat" may be to some extent imitative, it
is very doubtful if a person (or a chimp) not familar with the ASL would

be able to grasp the meaning of this hand signal. Likewise, most of the ASL, although slightly imitative, must be considered symbolic. However, as in all languages, there are a few ASL signals that will be classified as gestures. For example, the ASL signal for "drink" is the thumb extended from the fisted hand and touching the mouth.

The Gardners were able to shape some of Washoe's own gestures into signals that were good approximations of the ASL signals. For example, Washoe's begging gesture (extended open hand, palm up, towards a person) was successfully modified to approximate the ASL signal for "give me" which is similar to the chimp begging gesture except that it includes a prominent beckoning movement.

Both Sarah and Washoe were able to transfer spontaneously the signals from the specific object used in the initial training, to new members in each class of objects. For example, Washoe was originally taught the ASL signal of "open" on only three doors, however after Washoe learned this signal, she transferred this signal to all doors and then to other objects such as drawers, boxes, jars and even water faucets.

Also both Sarah and Washoe were able to combine some of their signals into simple sentences, such as "gimme tickle" or "please open hurry" (Washoe) or "Mary give apple (to) Sarah". Whether or not these are true sentences is open to debate. Remember that Chomsky (1965 & 1968) has argued it is the nature of our sentences which separates human from animal language. The evidence is unclear, but Terrace, Petitto, Sanders & Bever (1979) after analyzing more than 20,000 of these ape "sentences", concluded that these are not true sentences.

Fouts is now teaching seven new chimps the ASL hoping to see chimps communicate with each other in ASL. The new chimps are only being taught 36 signals, and Fouts hopes that Washoe will teach some of her additional 100 signals to them. We will have to wait and see if Washoe does this or not however.

These experiments have shown that a chimp's ability in using hand signals is truly amazing. Even with all these accomplishments however, chimps have still not bridged the barrier of symbolic language. These chimps have been able to learn some symbols we have given them, but have not yet produced a single symbol of their own. When Washoe was shown an object for which she had not been given a symbol, she would either mislabel it or else combine old symbols to describe it. For example, Washoe had only seen another chimp once since she was captured and had not been taught a symbol for "chimpanzee". When she saw another chimp, she labelled him "bug". When Washoe was first

shown a duck, she called it a "water bird". She was always limited to
the vocabulary we had given her, she was never able to invent a new
symbol of her own. She has never been able to symbolize.

True, many of the new symbols we humans invent are simply combinations of old symbols. Some examples are "television" and "astronaut". However, we often do this out of convenience, rather than out
of necessity. Our languages already have so many thousands of words,
that it is simply easier to remember new words if they are formulated
out of prefixes, suffixes and root words already familar to us. When
the machine was first invented, it was much easier for people to remember the word "television", than it would have been to remember a
completely new symbol such as "biaram". The important question is
how did we acquire these prefixes, suffixes and root words in the first
place? We acquired them from our human ancestors who created them.
Television is a combination of two older symbols *first created by
humans*. Chimps on the other hand, did not create either "water" or
"bird". Both these symbols were formulated by humans and given to
Washoe. Washoe was able to combine two symbols that we had given
her, but neither she, nor any other chimp, have ever created any symbols of their own.

Fouts has suggested that chimps might be able to invent their own
signals. He gives the example of when one chimp spontaneously bent
her index finger into a hook shape and touched it to her neck. Fouts
had not taught her any such signal. It was then realized that the chimp
wanted to go out for a walk, and the bent finger was referring to her
leash. True, the chimp had invented a new signal, but as should be
evident to anyone who has read this chapter, this signal was a gesture,
not a symbol. The bent finger touching the neck was imitative of the
leash. Chimps are able to create their own gestures, but not able to
create their own symbols.

If this chimp had instead, for example, patted the top of her head
when she wanted to go out, this would have been a symbol. There is
nothing about patting the top of the head that is suggestive of going for
a walk. Of course, if a chimp ever did create a new symbol, she
would have to teach it to us by repeatedly pairing it with the event, until *we* learned the connection. But they never have, nor do I think they
ever will.

EARLY LANGUAGE AND WRITING
     When our ancestors first acquired the use of language is not known.
Unlike stone tools, language leaves no material traces behind. Sym-

bolic language could only begin after man had acquired the ability to symbolize or create symbols. The ability to learn someone else's symbols (such as today's chimps have) is of no use in starting a language. A language cannot begin until someone with the ability to symbolize sets about creating symbols. Presumably chimps and many other animals have had the ability to learn someone else's symbols for thousands of years--but this ability by itself has never been enough for them to start their own symbolic languages. Unlike Sarah and Washoe, early man had no pre-formulated symbols given to him. Before man, there were no symbols. Early man had to be able to symbolize before any symbolic languages could begin.

It is generally assumed that the ability to symbolize was acquired by man very early--at least as early, and possibly earlier, than the first crude stone tools. One reason for making this assumption is that careful study of the so-called "primitive" cultures reveals that not only do all cultures possess symbolic language, but also that there is no society with a primitive or immature language. Of course, languages from these societies differ considerably from our own, but analysis of their grammar, structure and function shows that they are in no way inferior to ours. As a matter of fact, many of these languages are a good deal more complicated than English. For example, on Bougainville in the Solomons, the language has 20 different genders! (Latin and German have three genders, English none). Actually, linguists now argue that languages tend to develop in the direction of greater simplicity.

Many of the languages of the people of "primitive" cultures have fewer words in their language than English, but it must be realized that our inventions and borrowed words from other languages have padded our dictionary considerably. It can be shown moreover, that all languages have as many words as are needed by that society to express itself. If additional words become necessary in any society, either new symbols are created, or else symbols are borrowed from other languages to fill the need.

Another reason for assuming the great antiquity of symbolic languages is that we now know that the amount of cultural information possible to communicate without symbolic language is very limited. Without language, early man would have been limited to imitative learning and gestures in transferring acquired information on to others. Without symbolic language, it would have been very difficult if not impossible to pass on very early cultural improvements such as in stone tools. About one million years ago we find that early man's tool-making techniques were slowly improving--the passing on of these improved techniques would have been very difficult with only imitative learning at man's disposal.

We do have some idea when writing started however. The earliest writing of any sort appears around 6,000 years ago. Actually, early writing probably grew out of two-dimensional art, rather than as an extension of spoken language. As has already been pointed out, art like language is a form of communication. A drawing or a painting can represent and recall a particular event. Early writing most likely grew out of attempts to record events in drawings. All early writings were pictographic. That is, they were to some extent a picture of what they referred to. These pictographs were not related specifically to any spoken words. For example, the pictograph Λ , used by the Plains Indians, represents a dwelling, but does not refer to any specific Indian word. The Λ can be read as tent, tepee, dwelling, house, etc. There is no direct connection between pictographic writing and the spoken language. Pictographic writing is in essence a highly conventionalized form of two-dimensional art, having no direct relationship with spoken language.

Often as time went on, pictoraphs became simplified and abbreviated to the point where the original object was barely, if at all, recognizable. Also, additional characters were added to represent ideas that could not be represented pictorically. These ideograms, as we call them, still did not link up with any single word, however. Our mathematical symbols such as +, -, x, are examples of ideograms.

One distinct disadvantage of pictographic writing was that it required literally hundreds of different characters. Each object or idea required a different character, and all these had to be memorized. It required many years of training before one could read and write and as a consequence, very few people in any society were able to acquire this ability.

Writing as we now know it, began when these characters began to be associated with the sounds of spoken language. The passing of writing from pictographic form to our modern alphabetic form is quite complicated. The change over did not happen by any means all at once. Even when some of the early cultures such as the Sumerians and Egyptians first began to link up their characters with spoken language sounds, they still continued to use the pictographic characters as well. In the earliest sound writings, most characters represent spoken words; in other forms of early writing, characters stand for syllables and in our alphabetic writing, each character represents distinct sounds or phonemes. Most modern systems are alphabetical, although Chinese characters represent words and many Japanese characters represent syllables.

Our modern writing is a set of written symbols representing a set of spoken symbols. Our writing is a language of a language. Actually, our written letters represent only spoken sounds and nothing more. Properly speaking, the written symbol "cat" does not refer to a real cat, it only refers to the symbolic sound we make when referring to a real cat.

Written "cat" ⟶ Spoken "cat" ⟶ Real cat ⟵ (Pictographic cat)

As the diagram shows, the connection between our written symbol "cat" and the real cat is only indirect. A pictographic cat on the other hand refers directly to a real cat and has no connection with the spoken language.

One consequence of this is that although we may have a few spoken words that are imitative of the original event (such as "hiss"), we have no written words that are imitative. The spoken word "hiss" may be imitative, but the written word "hiss" is not. Show a Chinese the written word "hiss" and he will have no idea what it means. Early pictographic writing however, is to some extent imitative of the original object. As pictographic writing became more and more conventionalized (less pictorial) however, it gradually became less imitative.

Writing then, has a separate origin from that of spoken language. Writing has its roots in two-dimensional art, rather than in language. The invention of writing marks a major step in human progress. History is usually accepted as beginning when humans began recording events in writing. Everything which occurred before the advent of writing is usually referred to as "pre-history". Our knowledge of human events during the historical period is much greater and more exact—complete with names, dates and specific events—than the vague, nameless events and peoples we have been able to reconstruct from our prehistoric past.

Of course, a certain amount of "historical" information can be transmitted by word of mouth in pre-literate societies. Through a society's oral tradition, past events are "recorded" in the memories of living individuals and passed on by word of mouth from generation to generation.

However, there are major problems to this oral transmission of "history". As Goody & Watt (1963) point out, individuals tend to remember what is important to them and tend to forget facts which are no longer of significance to them. Thus, in each generation, individual members will unconsciously shift and sort their oral traditon before passing it on to the next generation. Facts which are no longer of contemporary relevance may well be eliminated by the process of forgetting. Also, any inconsistencies between what was believed in the past and what is believed now are likely to be forgotten or transformed. Thus, oral "history" tends to be automatically and unconsciously adjusted to suit the current situation by each generation. The past exists only in terms of the present. Myths and legends are modified and changed slowly over the years to the point where the original meaning may be lost or changed.

Oral "history" may thus be quite distorted after it has passed through the memories of several generations. And the individuals in a pre-literate society are completely unaware that they are receiving a selective and continually "modernized" version of the past. But with the advent of writing, the past cannot be altered, transformed and selectively discarded in the same way. With the beginning of history proper, the permanently recorded facts and beliefs of our ancestors are preserved intact. Individuals in literate societies are forced to acknowledge that people in the past thought and acted differently from the people of today. And, according to Goody & Watt, this recognition forced people in literate societies to be much more critical and sceptical of their own presently held beliefs. After the advent of writing, it was, for example, no longer possible to accept that present beliefs are eternal truths which were given to us by our ancestors.

The invention of writing allowed the transmission of human knowledge and the awareness of our past to reach new levels. Certainly spoken language was the first and most important ability that allowed acquired information to be passed on to other people. But with the addition of writing, the exact words, thoughts and records of our greatest leaders, thinkers and teachers were able to be passed on unchanged, not just to a few people, but to millions of people the world over and to countless generations yet unborn.

CONCLUSIONS

This concludes the review of the various abilities that have been proposed at one time or another to mark man apart from all the other animals. We have seen that most of these abilities have not stood the test of experimental scientific investigation. The apes, and the chimpanzees in particular, have proved to be the single most persistent challengers to our suggested uniqueness. Time and time again the chimp has been able to perform activities (admittedly on a simple level) that were formerly thought to be exclusively human. Even given that man is head and shoulders above all the rest of the animals, a strong case could also be made that the chimpanzee is head and shoulders above all the other non-human animals. The chimp is the only non-human animal that has an inkling of humour, the ability to recognize itself in a mirror, use of different tools for different purposes, and simple toolmaking. The chimpanzee is a very remarkable animal in its own right.

However, this book is about human intelligence, not chimp intelligence. My survey of abilities proposed as uniquely human has yielded only two main activities that we can still claim to be indicative of our mental uniqueness--art and language. The only remaining remnant of the once famous tool-use proposal, is the "ability to use one tool to make another tool". But by far, the proposed abilities of highest standing and importance today are symbolic language and art.

As noted previously, art and language are closely related. Both are forms of communication. Both forms of behaviour are learned or acquired. All cultures have symbolic language and also art. No culture can be said to have a primitive or immature form of either language or aesthetic art. Last and most important, both language and art exist solely on the basis of symbols or conventions. Neither form of behaviour can exist until someone proposes that one event shall represent another event. In language we call this symbolizing. When someone proposed that the word "hippy" shall mean a human with long hair and unconventional appearance, this is an act of symbolizing. In art, when someone decided that two-dimensional drawings would represent three-dimensional events, we call this forming a convention. Actually, both the act of creating a symbol and the act of creating a convention in art are the same. In both cases, we have assigned a connection or relationship between two dissimilar events, which had not been paired together previously. Language and art rest on the same ability. Any animal that was able to symbolize language would also have the mental ability to produce representational art.

More evidence that language and art are closely related comes from the history of writing. Writing began as a form of art, but slowly moved closer and closer to language. Writing now exclusively represents spoken language.

Desmond Morris makes an interesting argument about art, writing

and photography in his book *The Biology of Art* (1962). Morris states
that art has changed in nature and function as man has invented other
forms of two-dimensional communications. Before writing, art was
the only medium of preserving and storing information. Information
about an important event could only be preserved for posterity by
drawings or paintings or passed by word of mouth.

As writing evolved and became separate from drawing, art lost one
of its major functions. As a result of losing the function of preserving
historical information, art developed new methods of becoming increasingly representational or imitational of the original event. Artists developed several new techniques for producing pictures that were more
and more realistic or imitational of what they were painting. For example, the technique of optical perspective in art did not develop until
the 15th century in Florence, Italy. Until then, paintings had been primarily "conceptual"--that is, the most important subject was given
prominence even if this distorted the imitativeness of the picture. With
the technique of perspective however, paintings appeared to have three-
dimensional depth--the size of an object depended upon the distance and
not the object's importance. Art became more and more a detailed and
accurate visual representation or two-dimensional reproduction.

With the advent of photography, art suffered a loss of another of its
functions. Art is no match for modern colour photography for representing accurately every visual detail of a given event. First writing
and then photography robbed art of two of its original functions. Today
art is left largely with only the function of aesthetics--what is pleasing
and enjoyable, without also attempting to communicate any specific information. It could be argued that the advent and growth of the "abstract" school of art closely parallelled and in some sense was a reaction to the invention and development of photography. As photography
progressively took over more and more the function of detailed and
accurate visual representation, art increasingly became more aesthetic and less representational.

So art and language are really two aspects of the same ability.
Without the ability to symbolize, neither of the two could exist. Chimps
have not been able to create their own symbolic words nor have they
been able to develop their own representational art.

Chimps have been able to learn some of our symbols. Likewise,
if someone repeatedly paired a painting of a flower with a real flower,
it seems reasonable that the chimp could learn that a connection existed between the two.

The question of how chimps are able to interpret photographs is more difficult. Chimps are sometimes able to identify photographs of many different objects. However, it appears that at least at first, they are reacting to these photographs as if they were the real object rather than as a representation of the object. On exactly what level chimps are "interpreting" these photographs and whether they are ever able to recognize them as only two-dimensional representations is not clear. Very little experimental work has been done in this area.

A photograph is an extreme form of imitative two-dimensional representation. Few if any paintings and no drawings are as "life like" or realistic as a photograph. The degree of imitation in art can vary tremendously. A child's simple pencil drawing of a cat is far from being as imitative as a life-sized colour photograph of a cat, and there are many steps of imitativeness in between these two extremes.

Recall that a gesture was defined as a signal that was similar enough to the original event to allow comprehension *without previous training.* It is possible that chimps were able to "interpret" these photographs because the imitativeness had reached a point where they became a "pictorial gesture" rather than a symbol. The fact that they at first regarded the photographs as if they were real objects would support this possibility.

We have been exposed to paintings, drawings and photographs from such an early age, it is difficult for us to realize that these forms of art are usually symbols--that they are usually *not* similar enough to the original object to allow comprehension without training. But remember that some cultures do not have two-dimensional art, and the people of these cultures are not able to recognize correctly black and white photographs at first. These people tended to react either with total incomprehension, or else to regard the picture as the real object. Apparently once a picture had reached a certain level of imitativeness it became a "pictorial gesture" and like the chimps, they regarded the picture as the real object. It is likely that all these people and chimps as well would "recognize" a life-sized, colour photograph at first glance, however. They would "recognize" it not as a two-dimensional representation, but rather as if it were the real object. Anyone who has been temporarily deceived by a life-sized, colour store display photograph of a person will understand how it is possible.

The exact nature of interpretation of photographs and drawings by chimps and people not familar with two-dimensional representations is not yet clear. More comparative scientific work between man and apes is needed in this area, especially while there may still be a very few tribes of people left in the world isolated enough from our Western cul-

ture not to be familar with two-dimensional representations.
   If humans require training before we are able to recognize paintings
and drawings, we should not expect apes to recognize them without
training. Whether chimps could ever learn the more general convention
of representing three-dimensional space by a two-dimensional drawing
is not known. To my knowledge, no experimental studies of this type
have yet been performed.

   These questions of the ability of apes to learn to interpret photographs and paintings are all very interesting, but they are not after all
of primary concern here. On what level chimps are able to interpret
photographs or whether chimps will be able to learn to interpret some
of our artistic conventions are not crucial factors in this context. That
chimps can learn some of our symbols has been admitted. The important point is that *they cannot create their own symbols*. They cannot
create their own linguistic symbols nor their own artistic symbols.
Their own hand signals never develop beyond gestures and their "drawings" never develop beyond scribbles. This is the important point, and
to what degree or level they are able to learn our symbols is of secondary importance here.

   This is a genuine difference of kind. The ability to symbolize
marks us apart from the rest of the animal kingdom. And moreover,
this ability to symbolize has been very important to Homo sapiens.
Nearly all of our vast cultural transmission of information depends on
spoken language, writing and art. Certainly human life as we know it
would not be possible without the ability to symbolize.

   Virtually all the scientists who believe that man is different in kind
from the other animals believe that this difference is due solely to our
ability to symbolize. Leslie White has said "All human behaviour
originates in the use of symbols . . . It is the symbol which transforms
an infant of Homo sapiens into a human being" (1949, p. 22). It has
been widely assumed that a human child is no more intelligent than
other animals until it acquires speech and is then able to assimilate our
vast cultural heritage.

   With the great concern and enthusiasm for defending human language
as being a difference of kind, the men of the "kind" side have overlooked
another unique ability of humans, however. This ability has been hither
to unrecognized (or at very best only dimly recognized) up to now. It
is this second unique ability of man that will now be examined in Part
Two.

PART TWO

Chapter 3

Contiguous Learning

Humans have recognized from the earliest times that other animals, as well as man himself, are able to learn from experience. Learning is very widespread among nearly all, if not all species in the animal kingdom. It may very well be that learning is universal among animals, however, at present the evidence is not conclusive among the simplest of animals such as the protoza, jelly-fish, hydra and starfish.

Learning has been recognized for a very long time, as the fables of Aesop reveal. The theoretical study of learning also has a long history. Ever since Aristotle, men have reflected on the nature of learning. This philosophical study of learning reached its heights in the philosophical school of English empiricism. These laws of association were developed by men such as Locke, Hobbes, Berkeley, Hume, Hartley and the Mills. These philosophers viewed "learning" as an activity which was carried out in our "minds" or brains. Learning occurred when *ideas* became associated in the mind. Association theory held that simple ideas became associated or tied together into more complex ideas in our minds.

During this century there has been a dramatic shift in the meaning of this word however. This subject is now studied mainly by psychologists, and modern psychologists typically define learning as "adaptive change in individual behaviour as the result of experience" (Thorpe 1963, p. 73). It is important to note the word "behaviour" in the above definition. Traditionally most learning psychologists have been emphatic that learning should be defined solely in terms of observable behaviour. Even though an animal's ability to learn is obviously related to whatever processes are occurring inside his brain or mind, most psychologists have insisted that these mental processes should be deliberately ignored and that learning should be defined and studied in terms of the outward, objective, observable activities or behaviour of the animals. When an animal has changed its *behaviour* towards an object as a result of experience or practice, learning is said to have occurred. Whereas the philosophers had thought of learning in terms of *ideas*, most modern psychologists consider learning in terms of *behaviour*.

This modern concern with defining learning in terms of behaviour can best be understood in its historical context. When psychology first separated from philosophy in the 1800's, this new subject was defined as the scientific study of the "mind". For example, Wilhelm Wundt who set up the world's first psychology laboratory in 1879, declared that the object of psychology was the introspective analysis of the contents of immediate experience, that is, psychology should study our thought process from the "inside", we should study what was happening inside our "mind". The first psychologists were all concerned with studying the inner experience of the mind. Even though they conducted some experiments, they nevertheless believed that introspection should be the main form of observation.

Because of this mentalistic outlook of early psychology, Immanuel Kant declared about 1800 that psychology could never be a true science. Science requires experimentation and measurement of the objective world and it is impossible to measure or quantify the subjective phenomena in psychology. However, beginning in the first part of this century, psychologists began attempting to prove Kant wrong by redefining psychology's area of study in such a way as to make psychology into a "hard" or natural science. By noting that the natural sciences studied the objective external world, psychologists gradually abandoned their study of the "mind" and have progressively decided that their proper area of study should be "behaviour", that is the external observable manifestions of whatever was happening in the mind. We will become a natural science by studying the same types of things as the natural scientists study, they declared. Watson (1913) was the first to clearly state this view.

This "behaviouristic" definition of psychology has been accepted, to a greater or lesser extent, by most psychologists today. And the stronghold of behaviourism has always been in the field of learning psychology. This sub-field seemed to be perfectly suited for the behaviouristic outlook. Whereas the association philosophers and the mentalistic psychologists had thought of learning in terms of ideas, the early behaviourists deliberately ignored whatever mental events were occurring on the "inside" and concentrated on the objective manifestations on the "outside" (behaviour).

Psychologists began their experimental study of learning in 1898 with the work of Thorndike in America and in 1903 by Pavlov in Rus-

sia. Starting with these two scientific giants, interest in learning psychology has steadily increased, especially in America. Today it can
be said that the study of learning towers over all other fields in American psychology.

TYPES OF LEARNING

This behaviouristic view of learning has been amazingly successful
and has produced a wealth of solid research. The intense experimental study of learning during this century has revealed several types or
levels of learning. Below is a very brief review of the main types of
learning that have so far been recognized: habituation, classical conditioning, instrumental conditioning, insight learning, imitation and
"cognitive" learning.

HABITUATION Habituation is considered to be the simplest type of
learning. It is universal in all animals, even the protozoa and lower
metazoa. Although habituation is the simplest type of learning, it was
not the first type to be clearly recognized. Humphrey in 1933 was the
first to discuss clearly the nature of habituation.

Habituation is a learning not to respond to stimuli which tend to be
without significance in the life of the animal. More simply, it is a
decline of a response upon repeated stimulation. A good example involves the garden snail. Snails have a natural reaction of withdrawing
into their shells whenever a strange or suddenly moving object appears
near them. We might call these events "danger signals" to the snail.
If a snail is placed on a table and the table is hit with a hammer, the
snail will immediately withdraw into its shell. If this procedure is
repeated at regular intervals every few seconds, this withdrawal behaviour will fade until at last the snail will take no notice of our table
banging. The withdrawal behaviour can then be said to have become
habituated to this event.

Most animals (in particular small animals) have instinctive responses of withdrawal or self-protective action to certain classes of
stimuli. Thorpe (1963) has listed these as (1) any moving object (2)
any strange situation (3) any sudden stimulus or (4) any stimulus of
high intensity. These instinctive reactions help the animal avoid most
of the usual risks of life, but obviously it would be almost impossible
for an animal to respond to all the stimuli that fell into one of the
above four categories. Therefore there is a need for some form of
learning which modifies or eliminates the instinctive reactions to
those stimuli which experience has shown to be harmless. Habituation

then, is a learning not to respond instinctively to certain stimuli.

CLASSICAL CONDITIONING Ivan Pavlov (1849-1936) was the man responsible for first studying the conditioned reflex. Pavlov was a distinguished Russian physiologist before he ever began his work on learning. It was Pavlov's work on the physiology of digestion that first brought him fame and for which he was awarded the Nobel Prize in 1904. He was 50 years old when he first began his work on conditioned reflexes in 1899.

In order to understand classical conditioning it is first necessary to understand the term *reflex*. Most animals (including humans) will respond automatically to certain stimuli without any previous learning. Examples of automatic responses or reflexes are the contraction of the pupil when light is shone into the eye, and salivation in the mouth when food is presented to the animal. These reflex actions are unlearned. Reflexes occur automatically without any previous learning.

During his work on digestion, Pavlov noticed that sometimes the flow of saliva in the mouths of his dogs occurred before it was supposed to! The dogs began to salivate *before* the food was presented to them! How were the dogs able to do this? Pavlov discovered that his dogs were able to learn to associate food with the experimental setting.

This led Pavlov into his now famous experiments with the dogs and the tuning forks. First he presented the dog with the sound of a tuning fork alone to make certain that it did not elicit salivation before training. Usually the dog's only response was a turning of the head

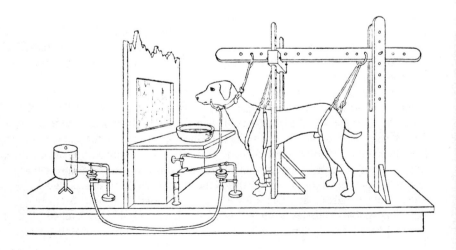

and cocking of the ears. Next, Pavlov presented the sound of the tuning fork just before or at the same time the dog was presented with the food. This pairing of the two was repeated several times. Then the test was run, Pavlov presented the sound alone, without the food. Sure enough the dog salivated to only the sound of the tuning fork. Before the training, the dog did not salivate to the sound alone, but now it did. The dog had learned to associate the sound with the food.

Pavlov called this learning the *conditional reflex* because the stimulus that produced the reflex was not the original one, but rather a new one conditional upon previous training. Therefore the new stimulus presented (in this case the sound of the tuning fork) bécame the conditional stimuli. Unfortunately the English translation of Pavlov's work improperly rendered what should have been condition*al* as condition*ed* and to this day it is called the conditioned reflex or the conditioned response.

Pavlov performed many more experiments dealing with the conditioned reflex. He carefully investigated many of the numerous empirical relationships of this type of learning. For example, he studied the extinction of conditioned relexes. Extinction is the loss of an acquired reponse when the response is no longer rewarded. In the experiment with the tuning forks and the dogs for instance, if the sound was repeatedly presented without the food accompanying it, the salivation response would gradually cease. Pavlov experimentally explored many other such relationships about the conditional reflex, but it is not necessary to discuss them here.

One aspect of the conditioned reflex that is important for us to understand, however, has been studied more recently by other scientists who are continuing work on this type of learning. This has to do with the amount of time that passes between the presentation of the new or conditioned stimulus (the sound in our example) and the original or "unconditioned" stimulus (the food). It has been found that in order for learning to occur, the two stimuli must occur close together in time, that is they must be contiguous in time. At most, not more than a few seconds can separate them if any learning at all is to occur.

One might think that the highest rate of learning would occur if the two stimuli were presented at exactly the same time, but this is not the case. It turns out that the optimum interval for learning varies between $\frac{1}{2}$ second and 10 seconds depending upon the response system being considered. If the interval between the two stimuli is increased beyond this, the rate of learning falls off sharply.

Early psychologists believed that if the events were presented together contiguously, learning would automatically occur. However, this view is now thought to be too simplistic as several recent experiments have shown that in certain situations mere contiguity is not sufficient to ensure conditioning. For example Rescorla (1968 & 1975) conducted an experiment in which he paired contiguously a tone and an electric shock. However, he also delivered "extra" shocks at other times. He found that if these extra shocks occurred frequently enough, they disrupted the tone-shock learning. Note that although the tone was always paired together contiguously with a shock, the extra shocks meant that the tone was useless as a *predictor* of shock. In other words, the animals are more sophisticated than we previously realized. The animals require that the contiguous pairings provide information about what is likely to happen in the future before they will display learning.

Classical conditioning then, is in a sense an alteration in reflexes; the animal learns to respond to a new or conditioned stimulus. Like the withdrawal reactions discussed in the habituation section, reflexes are innate, they are not learned. Whereas habituation is the learning not to respond to a certain stimulus, classical conditioning is the learning to respond to a new, additional stimulus. A review of the recent work done in this area can be found in Dickinson & MacKintosh (1978).

INSTRUMENTAL CONDITIONING OR TRIAL AND ERROR LEARNING
The two types of learning previously discussed were concerned with the instinctive or reflexive responses of animals. However, much animal behaviour is not of this nature. Particularly in the higher animals, a very large proportion of their responses are said to be spontaneous, "voluntary" and flexible. The higher in the animal hierarchy one looks, the more important this "voluntary" and pliable behaviour becomes. Nearly all human behaviour is of this type. It is these flexible responses that are involved in instrumental conditioning.

The first man to experimentally study this type of learning in animals was Edward L. Thorndike (1874-1949). His *Animal Intelligence* was written as his doctoral dissertation in 1898 when he was only 24. Thorndike's first experiments were done mainly with cats. In the typical experiment, a cat (that had not been fed for several hours) was confined in a cage with a door which could be opened from the inside by pulling on a wire loop. Food was placed outside the cage. Note that part of the door opening mechanism was concealed from the cat's view. Thorndike then timed the cat to see how long it took him to get

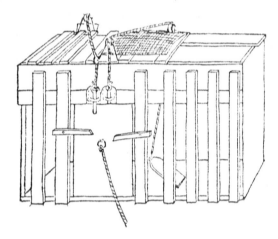

out. On the first attempt, each cat made many ineffective movements--
clawing, biting and dashing about, but finally the cat's varied and random activities led it to pull the loop. When the loop was pulled, the
door was pulled open and the cat could escape to reach the food. On
succeeding attempts, the time it took the cat to escape became gradually less, but this decrease was slow and irregular (see graph).

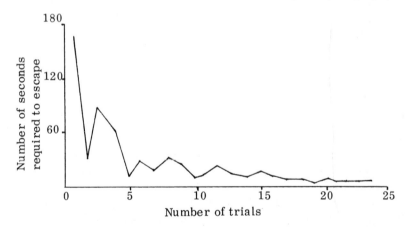

Eventually, after many repetitions, the cat was able to pull the loop
almost as soon as he was placed in the cage.

Thorndike called this type of learning "trial and error" learning.
He developed a theory that once the animal hit upon the successful
movement (that is, a movement which resulted in obtaining food), the

animal gradually learned to associate this movement with the opening
of the door. This association was called a "connection" or "bond".
Thorndike believed that this connection was gradually strengthened
over the successive trials. The greater the number of successful
trials, the stronger the connection (between pulling the loop and the
opening of the door) became. His famous "law of effect" refers to
this strengthening or weakening of a connection as a result of its consequence.

Note that when first placed in the cage, the cat produced a wide
range of different behaviour (scratching, biting etc.). These responses
were certainly not reflexive or fixed instinctive reactions. These
were the so-called "voluntary" and flexible responses that characterize
so much of mammalian behaviour. Through "trial and error" the animal finally stumbled upon a successful response. It was the very flexibility of the animal's responses that enabled it finally to escape.

Although Pavlov (1903) and Thorndike (1898) were both "behaviourists" in the sense that they studied learning in animals by observing
changes in their behaviour, it seems likely that they were behaviourists out of necessity rather than out of choice. Since they only studied
learning in *animals*, they almost had to be behaviourists, since it was
impossible to talk to animals and ask them what mental processes
were occurring in their heads. The first clear theoretical proclama-
tion of behaviourism was not until Watson (1913). Watson proposed
that psychologists should be behaviourists on principle rather than just
out of necessity when studying other animals. To this day however,
behaviourists have done most of their experimental work on other animals (and in particular the rat), perhaps because when studying other
animals, psychologists cannot be tempted back into a mentalistic point
of view.

A more recent development in the study of instrumental learning
has been the "Skinner box" after B. F. Skinner (1938) who invented it.
It is a small box which contains a small lever, a food cup and a light.
When the lever is pressed a small pellet of food may fall into the cup.
The lever is also connected to a recording pen, so that a record is
automatically kept of each time the lever is pressed. Usually it is the
laboratory rat that is placed in the Skinner box. The box is thus an
ideal machine for studying instrumental learning in rats and other
small animals.

Like the cats in Thorndike's experiments, when first placed in the
box, the rat displays a wide range of flexible and "exploratory" behaviour. Eventually it will hit the lever and a pellet drops. Every time

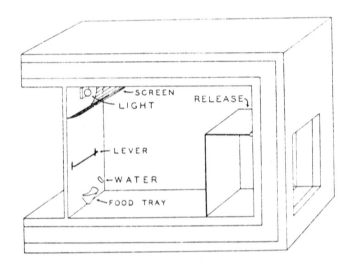

the rat is rewarded or reinforced with food, after pressing the lever, this strengthens the lever pressing response. In other words, every time the rat obtains food after pressing the lever, the more likely it is he will press the lever again. Perhaps a similar behaviour would occur in humans if, say, a child accidentally found that by pressing a small button on the side of a sweet machine, he received a free choc- olate bar. Each time he was "rewarded" after pressing the button the more likely he would press the button again--at least until he had eaten his fill.

There have been literally thousands of experiments conducted using the Skinner box. All sorts of variables have been examined. For ex- ample, the rat can be rewarded every time it presses the lever, or it can be rewarded only every third, or tenth or hundredth time it pres- ses the lever. Also, the rat can be rewarded only every three min- utes or every ten minutes and so on. When the animal is rewarded only part of the time, this is called partial or intermittent reinforce- ment. The occasional reward given by slot machines is an example of intermittent reinforcement.

Skinner's explanation of how learning occurs differs somewhat from Thorndike's. However, it is not necessary to explain those differences here. What is important to understand is that like classical condi- tioning, instrumental learning must be contiguous in time. In order for learning to occur, the pellet must drop within a few seconds after the animal has pressed the lever. The two events must be contiguous. (An "event" can be either a stimulus or a response.) If the interval between the two events is more than 2-5 seconds, the rate of learning decreases significantly. Apparently no learning at all is possible if the interval is greater than about 30 seconds, but this matter is somewhat complicated and will be discussed in more detail in this chapter.

As Tolman (1932) and Thorpe (1963) have pointed out, instrumental conditioning--similar to classical conditioning--can be regarded as expectant in nature. After the rat has learned the connection between the lever pressing and the dropping of the pellet, it certainly acts as if it *expects* the food to follow the lever pressing. As a result of the learning, the animal seems to have a certain amount of foresight regarding the consequences of pressing the lever.

Instrumental conditioning is an infinitely more important type of learning than either classical conditioning or habituation. It is this trial and error learning that adjusts the "voluntary" actions of an animal to its environment. Habituation and classical conditioning only involve the relatively fixed reflexes and instinctive reactions, whereas trial and error learning encompasses the vastly more important voluntary responses. A very large part of human learning is of this trial and error type.

GESTALT INSIGHT LEARNING Although listed here as a separate type of learning, in some respects this is not an entirely accurate classification of "insight learning". Instead of being thought of as a separate type of learning, in many ways it is rather an alternative explanation for instrumental or trial and error learning.

It was Wolfgang Kohler (1887-1967) who first coined the term "insight" learning in his classic study of chimpanzee problem solving titled the *Mentality of Apes* (1927). Detained during World War I on a small island off the coast of Africa, Kohler, a German gestalt psychologist, was able to study a group of native chimps.

Kohler's experimental set up differed considerably from those of Thorndike and Skinner. Kohler tested the chimps under a wide range of situations which always provided the opportunity for the chimps to solve the problem in an intelligent manner using "insight". Unlike Thorndike or Skinner, the experimental situations used by Kohler were always fully exposed to the animal. If the animal was smart enough, he could "see" the solution to the problem.

The gestalt school never completely accepted the behaviouristic demand that the internal subjective experience of the individual should

be completely ignored. The gestalt theorists believed that an important aspect of learning was how the animal *perceived* the world, rather
than the objective world itself. Thus, although they were behaviourists in the sense that they first studied the external behaviour of the
animals, then, unlike the early or radical behaviourists, they believed
it was permissible to hypothesize or theorize what was happening on
the "inside" or in the animal's mind. As a consequence, they have
sometimes been referred to as "neobehaviourists". The gestalt psychologists have tended to study learning in the more intelligent animals and their outlook is usually regarded as more representative of
the "higher" learning in humans.

A good example of the many experiments run by Kohler is as follows: the chimp is inside the cage with only a short stick at his disposal. Outside the cage lies the fruit and a longer stick. Neither the

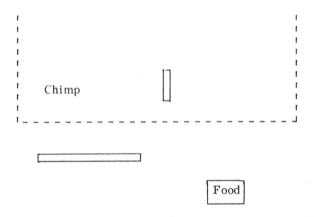

food nor the long stick can be grasped by hand. The chimp has previously learned to pull in food out of his arm's reach by using a stick.
In this experiment however, the food can not be reached by using the
short stick, but the longer stick is within range of the short stick.

Sultan /the most intelligent of Kohler's chimps/ tries to
reach the fruit with the smaller of the two sticks. Not succeeding, he tears at a piece of wire that projects from the
netting of his cage, but that, too, is in vain. Then he gazes
about him; (there are always in the course of these tests some
long pauses, during which the animals scrutinize the whole
visible area). He suddenly picks up the little stick once more,
goes up to the bars directly opposite to the long stick, scratches

it towards him with the "auxiliary," seizes it, and goes with
it to the point opposite the objective, which he secures. From
the moment that his eyes fall upon the long stick, his procedure
forms one consecutive whole, without hiatus, and, although
the angling of the bigger stick by means of the smaller is an
action that *could* be complete and distinct in itself, yet observation shows that it follows, quite suddenly, on an interval of
hesitation and doubt--staring about--which undoubtedly has a
relation to the *final* objective, and is immediately merged in
the final action of the attainment of this end goal. (1927, p.
174-175)

Note that all the necessary components of a solution to the problem
are clearly visible to Sultan. Remember that in Thorndike's experiments with cats part of the door release mechanism was concealed
from the cat's view. Also the Skinner box is constructed in such a
way that the animal cannot observe why pressing the lever causes the
food pellet to fall. Kohler critized Thorndike's tests because his experimental situation necessarily precluded any intelligent solution by
the animals.

From his experimental results, Thorndike had developed his theory
of trial and error learning. The animal seemed to begin with completely random movements and only accidentally stumbled upon the
correct movement. The correct connection was then slowly stamped
in during repeated successful trials. Kohler argued that this random
or "blind" trial and error occurs only when the animal is either not
sufficiently intelligent or else when all the necessary components of a
solution cannot be observed by the animal. An insight solution occurs
when the animal is able to apprehend the relationships of a problem
and obtain a reward without resorting to a "blind" trial and error
method of solution.

How is it possible to distinguish between blind trial and error and
insight solutions? There are at least four criteria for insight solutions. (1) There is an interruption of movement by the animal for a
short period during which the animal appears to be carefully inspecting
the situation. (2) This period is followed by the animal suddenly and
directly performing the necessary movement. There is no fumbling
about, no random or unnecessary movements. (3) The animal is able
to repeat the solution readily after this single critical solution. Recall in Thorndike's trial and error experiments, many trials were
necessary before the animal could readily respond correctly. (4) The
animal should be able to generalize the insight solution to new prob-

lems which employ the same principles or relationships.

Kohler and other gestalt psychologists hold that the more intelligent
an animal, the more likely it is to achieve an insight solution. Certainly humans have the highest capacity for insight learning. In man,
when someone finds an insight solution to a problem, we often refer to
this as an "Aha" experience. We are suddenly able to visualize clearly
the correct solution to the problem. Chimps appear to have the
greatest ability among non-human animals, but other animals such as
the dog, cat, guinea pig and even the rat are also capable of some insight learning.

Insight learning, like the previous forms of learning discussed,
must be contiguous. In order for an animal to achieve an insight solution, all the necessary parts need to be directly visible. If a needed
tool is hidden or not in the animal's perceptual field, the likelihood of
a correct solution is greatly reduced. If a stick is lying on the opposite side of the cage from the food, and hence the animal cannot see it
at the same time as the food, it is much harder for the animal to learn
to use the tool. This is the gestalt equivalent to the contiguous requirement in classical and instrumental learning.

For example, Kohler found that chimps were unable to use sticks
to draw food into their cages if

the experimenter takes care that the stick is not visible to the
animal when gazing directly at the objective--and that, vice
versa, a direct look at the stick excludes the whole region of
the objective from the field of vision--then, generally speaking
recourse to this instrument is either prevented or, at least,
greatly retarded, even when it has already been frequently
used. I have used every means at my disposal to attract
Tschgo's attention to the sticks in the background of her cage
and she did look straight at them; but, in doing so, she turned
her back on the objective, and so the sticks remained meaningless to her . . . /However/ sticks--and other substitutes--
which she beheld in the direction of her objective, were made
use of without any hesitation . . .

We subjected Koko to a similar test with similar results.
He made useless efforts to reach the objective: a stick was
quietly placed behind him; but though, on turning round, he
looked straight at the stick and walked across it, he did not
behold in it a possible implement. If the stick was silently
moved towards him, so that the slightest movement of the
head or eyes would lead from the region of the objective to

the stick--suddenly he would fix his gaze on it, and use it.
(p. 37-38).

It is probably incorrect to classify insight learning and trial and
error learning as two separate types of learning. It is most likely
best to think of them as two ends of a single continuum. Usually some
trial and error behaviour is present in achieving an insight solution.
Often an animal makes several false starts before finding the correct
insight solution.

Actually, an animal seldom produces truly random movements in a
trial and error situation. Usually a "try" by an animal can be considered to be a real try and not just any random action. Animals often
appear to by trying out various "hypotheses" in a problem setting,
first trying one approach, then another. Even psychologists who favour the trial and error interpretation of learning sometimes speak in
terms of approximation and correction.

Insight theory implies that an animal always acts as intelligently
as it can when confronted with a problem. If through a combination of
intelligence, previous experience with similar problems, and favourable experimental situation, the animal is able to perceive the essential relationships of the problem, then the animal is able to achieve an
insight solution. If however, the animal is not able immediately to
apprehend the solution, he must make various "trys" at the problem.
A strong case can be made that true random or blind trial and error
learning is not the pattern of all learning.

IMITATION This type of learning has already been briefly discussed
in Chapter 1. Although Thorndike (1898) devoted considerable attention to imitative learning, most modern learning psychologists have
all but ignored this mode of learning. Perhaps because imitation does
not fit into the general framework of associative learning, most of today's textbooks of learning mention imitation only briefly, if at all.

Imitative learning occurs when the learning animal observes another
animal performing a certain activity. The learning animal is then able
to perform this activity later. The learner is often able to repeat this
activity successfully on the first trial. When first trial learning occurs,
the learning occurs without reward (since the activity never occurred
before). Imitative learning does not fit into the framework of associative learning because it can often occur on the first trial (no previous
trial and error) and also because there has been no previous reward.

Animals that are able to learn by imitation have a tremendous advantage over animals that do not have this ability. Learning by imita-

tion eliminates the necessity of each individual learning by trial and
error. Once one animal discovers a useful activity, this knowledge is
passed on to other animals in the group. It is not necessary for each
animal to discover by trial and error each activity on his own.
Learning is shared. Learning becomes a group effort instead of an
individual effort.

Apparently, only primates and perhaps cats are capable of true imitative learning. Thorpe (1963) defines true imitation as copying a
novel or otherwise improbable act or utterance, or some act for which
there is clearly no instinctive tendency. Although birds and some
mammals display what may at first appear to be imitation, careful
analysis shows that these behaviours cannot be regarded as true imitation and are instead classified as either "social facilitation" or "local
enhancement" (see Thorpe, 1963, p. 132-135). Clear evidence of true
imitation is found only in monkeys, apes and of course man. Examples
of imitation in monkeys and apes have already been discussed in Chapter 1 (p. 37-38).

"COGNITIVE" LEARNING Humans, of course, are capable of all the
types of learning so far discussed. However, the most important
learning in humans is undoubtedly the "higher" type of learning similar to the gestalt insight learning. This study of the "higher" learning
in humans is usually now referred to as "cognitive" learning, although
it grew out of and still has some links with the earlier gestalt insight
outlook. The cognitive theorists, like the gestalt theorists, are not
strictly behaviourists in that they believe internal mental processes
should be considered. Thus, cognitive psychologists still largely view
learning from the "inside" and study such things as concept formation,
perception, memory, attention, problem solving and consciousness in
humans. A review of cognitive psychology can be found in Neisser
(1967 & 1976).

Thus in the field of learning psychology, the strict behaviourists
have done most of their work on non-human animals, whereas the study
of human learning has largely been left to the "cognitive" psychologists.
The important point here is that this field is thus split between the
study of learning in animals which is largely viewed from a behaviouristic point of view and the "higher" learning in humans which is still
largely viewed from a mentalistic or cognitive point of view. This
split is largely unconscious among today's learning psychologists, although this was not the case a few years ago when behaviourism and
mentalism were still battling for supremacy. For example Carr (1927),

in his presidential address to the American Psychological Association, admitted that he was a behaviourist when dealing with animals, but a mentalist when dealing with humans.

Human learning is also accepted as being limited to contiguous learning. The cognitive psychologists have not done much work in this area however. As we will see later, experiments carried out on human subjects have usually differed somewhat in design and purpose from experiments done on animals as a result of the differing outlook of the cognitive school.

THE PRINCIPLE OF CONTIGUITY

Learning psychologists inherited the principle of contiguity from their philosophical ancestors, the English empiricist philosophers, Locke, Hobbes, Berkeley, Hume, Hartley and the Mills who had in turn inherited it from Aristotle. Aristotle had stated three principles of association: contiguity, similarity and contrast and the English empiricists proposed these as well as other principles or laws of association. Each of these writers differed somewhat in the laws or principles they proposed, but the only principle which found its way on to every list was that of contiguity.

The association philosophers had talked of contiguity in terms of ideas--those ideas which occur close to each other in time tend to become associated. The behaviourist psychologists, of course, redefined this principle in terms of behaviour. Learning occurs when observable events and behaviour occur contiguously. They found that the stimulus (for example the appearance of food) and the animal's response must occur within a few seconds of each other for learning to occur. After the rat in the Skinner box has pressed the lever, the food must drop within a few seconds if learning is to occur. Their experiments convinced them that the stimulus and response *must* occur close together in time if any learning is to occur. All learning psychologists have accepted this principle, one theorist, Edwin Guthrie (1952) maintained that this was the *only* principle of learning. Even gestalt insight theory, which had been the most critical of the traditional associative interpretation of learning, agreed that all the components of the problem need to be visually present *at the same time* for insight learning to occur. There are only two narrow exceptions to the principle of contiguity which have recently been discovered, these exceptions will be discussed later.

An interesting set of experiments usually referred to as "delayed reaction" tests are sometimes confused with the contiguous learning

problem. The first delayed reaction experiments were conducted by
Hunter in 1913. In this type of test, the experimenter placed some
food in one of three containers while the animal is watching. The animal must then wait a given interval before he is allowed to approach
the three containers. The animal fails the test unless he goes directly
to the baited container without first trying any of the others. Various
species have been tested to find the longest delay they can accommodate and still remember the correct container. Of course, this is in
essence a test of memory. Apparently, memory varies from species
to species. For example, cats are able to make the correct choice
after a delay of 16 hours. Monkeys for at least 20 hours and chimps
for at least 48 hours.

These delayed reaction experiments should not be confused with
contiguity in learning. Remember that in the learning situation such
as the Skinner box, the animal presses the lever and then the food pellet drops. In order for learning to occur, the pellet must fall within
a few seconds after the lever has been pressed. This time interval
between the two events is usually referred to by psychologists as "delay of reinforcement". In the first set of experiments discussed, (the
"delayed reaction" experiments) the problem for the animal was only
one of memory, there was no learning required. In the Skinner box
however, the animal is required to learn that a connection exists between the pressing of the lever and the dropping of the food. Although
memory is obviously involved to some extent in this problem, there is
also an additional problem of learning that a connection exists between
two events.

Psychologists now accept that the maximum delay of reinforcement
interval is probably less than 30 seconds, or at the absolute maximum,
60 seconds if learning is to occur. The exact number of seconds that
can separate the response and the reward is still not agreed upon. In
order to understand how psychologists reached these figures, it will
be necessary to examine their experimental proceedures in more detail. A review of the work done in this area can be found in Renner
(1964).

A big part of the problem of discovering the maximum possible interval between response and reward lies with the phenomena of what
has come to be called "secondary learning". *Secondary learning*
occurs whenever any additional stimulus is consistently and closely
associated with the reward.

An example of secondary learning would be if, on your daily walk
through the park, you found some money on several different occasions

(spread over several weeks) under the same tree. After your first
experience in accidentally finding the money, you would probably give
at least a quick glance at that particular tree each day as you walked
by. After several rewards, you would probably carefully inspect all
round the tree each day. You would not know the "cause" of the money
being there. You would probably not think that the tree caused or produced the money*, nor would you think that your act of walking by the
tree in anyway produced the money. But nevertheless, you would realize that there was "something about that tree", that that particular
tree was in some way related to the money. The tree would be the
secondary stimulus. You would not know the primary event or "cause",
but you would be able to form a secondary connection with the tree.

The rat in his Skinner box is also capable of secondary learning.
In the Skinner box the lever is usually placed near the food tray.
Therefore whenever the food drops in the tray, the lever is always visually present. The rat is always able to see the lever at the same
time that the pellet appears. The sight of the lever becomes a secondary stimulus for the food, just as the sight of the tree became a secondary stumulus for your reward. The rat may not know that the
cause of the pellet appearing was his pressing the lever sometime ago,
but through secondary learning, he learns that, as we would put it,
"there is something about that lever". Once the lever becomes a secondary stimulus, the rat will probably spend a great deal of time near
the lever, inspecting it, and incidentally pressing it.

Note that secondary learning is also contiguous. The secondary
stimulus must be contiguously associated with the reward. Your
money was always contiguous with the tree. For the rat, the sight of
the lever must be contiguous with the reward. But since the lever is
always present, the animal is capable of showing some learning over
an extended interval between his lever pressing and the arrival of the
food. And of course in this situation, it is impossible for the psychologist to know if the rat is showing primary learning or if the pressing
of the lever is only a result of secondary learning.

Early experiments in the maximum delay of reinforcement interval
produced widely varying results of anywhere from 4 seconds to 20 minutes. We now know that this wide diversity of results was due mainly
to the secondary learning phenomena in some of the studies. The matter was greatly clarified by work of Perin (1943). He used a modified

---

*Unless you happen to live near a park where money really does grow
 on trees!

Skinner box to eliminate as much as possible any secondary learning. In Perin's study, after the rat pressed the lever, the lever immediately and silently withdrew into the wall and therefore when the pellet was dropped X number of seconds later, the lever was not present. Perin found that with the secondary learning phenomena all but eliminated, primary learning was possible with a maximum time interval of less than 30 seconds.

The maximum interval of primary learning is still not completely settled however. Spence (1947) has proposed that all learning occurring under delay conditions is possible only because of internal cues in the animal such as sensory traces which persist for several seconds, and it is this internal trace which is contiguous with the reward. According to Spence the maximum primary learning interval is not more than 5 seconds. Other experimenters have suggested 21, 34, 44 seconds as the maximum interval. Fortunately the exact number of seconds of the maximum interval is not of crucial importance here. The important point is that virtually all psychologists agree that primary learning cannot occur with intervals of more than 30 or at the very most 60 seconds.

Since the primary learning interval is so short, it is secondary learning that has been used to account for learning that occurs over several *minutes* rather than in *seconds*. For example, Pavlov reported in his classical learning experiments that sometimes his dogs would start salivating at the sound of the experimenter's footsteps approaching the laboratory. The sound of footsteps was of course heard several minutes before the dog received any meat. We now realize that this learning was only secondary learning--the experimenter was consistently present when the dog received the meat.

The maximum interval in minutes that secondary learning makes possible has not been determined. It varies greatly with the situation. As pointed out before, the learning of the secondary stimulus must be contiguous, just as with the primary stimulus. It is only the secondary stimulus that makes learning *appear* to be spread out over several minutes. Actually, all the animal's learning has still been restricted to an interval of 60 seconds or less.

The important point is that, without the benefit of secondary learning, psychologists have, until recently, universally agreed that all types of associative learning must be contiguous, that is within 60 seconds or less. There are only two recently recognized limited exceptions to this contiguity principle. The first of these exceptions will be discussed below.

Hypothetical ex-
ample of second-
ary learning by a
dog. The dog has
learned that riding
in the car is con-
sistently connected
with going to the
veterinary doctor.
Riding in the car
has become a sec-
ondary event.

"THE ONLY PLACE HE EVER *GOES* IS TO THE VET'S ....
THAT'S HOW HE KNOWS!"

Most of the experimental work in this area has been conducted with
animals and in particular rats. It has been generally accepted however that the principle of contiguity holds for humans as well. A few
studies have tested humans, but these studies differ somewhat in design in that nearly all of them use delayed feedback of how the person
performed on a simple skill. The person was asked to perform a simple task, then the experimenter waited a given interval before telling
the person how well he had done. Unfortunately neither the behaviouristic nor the cognitive theorists have bothered to carry out experiments with humans which duplicate exactly the experimental conditions of Perin when he attempted to eliminate secondary learning in
rats.

Why is learning limited to contiguous connections? We do not really know, but one plausible explanation has been put forth by Spence
(1956) and Revusky (1971 & 1977). They proposed that with longer

and longer intervals, the animal is likely to experience more and more events which will interfere with the response being learned. In other words, with a long interval, the animal will not know which of the several events that occurred recently is the "cause" of his reward. The longer the interval, the greater the number of competing or interfering events and the poorer the rat's performance.

The idea eventually led to the discovery of one of the two narrow exceptions to the principle of contiguity now recognized by psychologists. Revusky (1971) suggested that if extraneous events occurring during a delay interfere with learning, perhaps events which the animal experienced in a different situation would create less interference than events experienced in the same situation. In other words, if the animal was removed from the experimental situation immediately after he had made his response, and spent the delay interval elsewhere, perhaps there would be less interference than if the animal spent the delay in the experimental situation.

Lett (1973,1974 & 1975) tested Revusky's proposal. Lett (who, as it turns out, is Revusky's wife) removed the rats from the experimental situation during the delay. Lett's experimental procedure differed somewhat from that of Perin's, but it is not necessary to discuss these differences here. The important point is that like Perin, Lett carefully controlled for secondary learning. Immediately after the rat had made its response, it was picked up and placed in its home cage to spend the interval. At the end of the interval the rat was again picked up and returned to the experimental situation where it received its reward. Sure enough, Lett found that the animals were able to learn with intervals of at least 1 hour! These results appeared to provide strong support for Revusky's interference theory.

However, difficulties soon developed. First of all, Roberts (1976) was unable to replicate Lett's results. Then, Lieberman, McIntosh & Thomas (1979) and Lieberman (1980) did replicate Lett's findings, but by conducting several different experiments, found that the crucial factor which allowed the rats to learn over these long delays was not whether the animal was removed from the experimental situation during the delay, but rather whether the animal was picked up or not! They found that animals which were picked up immediately after their response, *but then replaced in the experimental situation*, were also able to learn over long delays.

Lieberman *et al* proposed that perhaps the picking up of the rats by humans was a particularly salient experience for the rat and this experience may have *marked* the response in some way. Perhaps it

was the marking of the response which allowed learning over long delays. To test this idea, they then repeated the same basic experiment, but this time instead of picking the rats up, they sounded a loud
noise or flashed a bright light immediately after the animal had made
his response. Learning with long delays still occurred.

It is still much too early to know exactly what is going on here, but
at present the evidence seems to indicate that if the cause is "marked"
with a strong salient event, learning with delays of over 1 minute is
possible. The important point however, is that all psychologists
would agree that if the animal is left in the experimental situation undisturbed during the interval (such as in Perin's experiment), learning
with delays of more than 60 seconds is not possible. It is only when
a strong prominent stimulus "marks" the event, that animals are capable of learning with a delay of over 60 seconds. There is only one
other narrow exception to the principle of contiguity and this has to do
with feeding and poisoning of rats and other animals and this exception
will be discussed in Chapter 6.

Thorndike (1931) believed that although humans could of course
form many more connections between events than other animals, there
was nothing qualitatively different about the connections we are able to
learn. Today psychologists generally agree with this view that although man can obviously learn a great deal more than other animals,
this difference is only a difference of degree. No modern learning
psychologist has ever proposed that our ability to learn is different in
kind from the learning ability in other animals. No psychologist has
ever proposed that the human species might be exempted from the
principle of contiguity. Our learning ability is supposed to differ only
in degree, not in kind.

There is another type of learning however, that is the exclusive
domain of humans and has not yet been recognized by psychologists.
This uniquely human form of learning will be explained in the next
chapter.

Chapter 4

Human Learning

Most school children are familiar with the names of Lister and Pasteur, but few people have ever heard of Ignaz Semmelweis. Semmelweis (1818-1865) was a Hungarian physician who did most of his important work in the Vienna General Hospital from 1844 to 1848. Although his work preceded that of Pasteur or Lister, his contribution to the understanding of infection and germ theory has never been widely publicized. Below is a brief account of Semmelweis's work on puerperal or childbed fever. Semmelweis's work will be used as an example not only in this chapter, but also later in the book, so it is important that the reader understand at least the general outline of his interesting work. A more complete account of his work can be found in Sinclair (1909).

IGNAZ SEMMELWEIS

The son of an affluent Budapest merchant, Semmelweis originally went to Vienna to study law, but changed to medicine after a chance visit to an anatomy demonstration. After receiving his M.D., Semmelweis was assigned to the maternity division of the Vienna General Hospital in April of 1844.

Childbed fever was a particularly dreaded disease in Europe at this time. Most hospitals in the major cities had repeated outbreaks of childbed fever. Although most women still delivered their babies at home, those women who, because of poverty, illegitimacy, or obstetrical complications were forced to seek hospitalization were faced with mortality rates occasionally ranging as high as 25%-30%. The first symptoms of the disease usually occurred two or three days after the woman's delivery. First the woman's pulse would rise, then raging fever and a discharge from the womb would develop. Delirium and diarrhoea usually followed and finally death.

Although the Vienna General Hospital had one of Europe's most enlightened medical schools, the incidence of this disease was as bad if not worse than in other hospitals. When Semmelweis joined the hospital he was already aware of its bad reputation for puerperal fever—in 1844 almost 10% of the pregnant women who entered the hospital died of childbed fever.

There were as many as thirty different conjectures or hypotheses popular at this time to explain the origin of the disease. It was said to depend upon the weather, or whether the mother was married or single, or on overcrowding in the hospital and so on. The most commonly accepted medical explanation was that it was caused by an invisible "miasm". This was a mysterious epidemic influence of an "atmospheric-cosmic-telluric" nature.

This was the situation when Semmelweis arrived at the Vienna General Hospital in 1844. In his search for the solution to childbed fever, Semmelweis was helped out by a chance reorganization of the maternity division four years earlier. The division had been divided into two clinics. The two different clinics were run in much the same way except that the first clinic was staffed by medical students and the second clinic by midwives.

Most of the women patients believed that the death-rate was higher in the first clinic than in the second, although this was discounted by most of the medical teaching staff. Admissions to the two clinics occurred on alternate days, Sunday for the first, Monday for the second, and so on. In attempting to be admitted to the second clinic, pregnant women would sometimes wait outside the hospital in horse-drawn buggies until just after midnight. Some women would rather give birth on the street than be admitted to the first clinic. One woman, when she found she had miscalculated and was to be admitted to the first clinic got down on her knees and begged Semmelweis to let her go home.

Semmelweis wondered whether the women might be correct that the first clinic was more dangerous than the second. Fortunately the hospital kept exact records of admissions, deaths and causes of deaths. Semmelweis checked, and sure enough since the two clinics had been established, about 10% of the women admitted to the first clinic had died of childbed fever, but only about 3% of the second clinic patients had died. All the doctors in the maternity division were aware of the patients' fear of this first clinic, and all of them had access to the records, but Semmelweis was the only one to check this possibility.

After Semmelweis had confirmed that there was indeed a difference between the two clinics, he realized that the explanation then accepted by most of the medical staff—the "atmospheric-cosmic-telluric" influences—could not account for these differences. How could this vague epidemic influence plague the first clinic but not the second, Semmelweis asked. He rejected this commonly accepted medical explanation completely and began to look for another answer. This rejection may well have been the most important step of Semmelweis's work; often it appears that the hardest part of a scientific breakthrough is rejecting the currently accepted scientific explanation.

Semmelweis already had set up for him what scientists today would call a control group. The two clinics had different rates of childbed fever, and Semmelweis correctly reasoned that there must be a reason for this difference. Although the two clinics were run along quite similar lines, there were some differences, and Semmelweis realized that the key to the understanding and prevention of childbed fever lay in one of these differences.

Semmelweis began to examine carefully the differences between the two clinics. Having totally rejected the currently accepted explanation, he was travelling in uncharted waters. He wrote, "Everything was questionable, everything inexplicable, only the great number of dead was an undoubted actuality." He began noting all differences between the two clinics, even those that seem obviously trivial or unimportant to us today. He then tested these differences one by one to see if this would effect the fever rate in the first clinic.

One difference that came to his attention was that the priest's route in bearing the last sacrament to a dying woman always passed through the first clinic, but not the second. Semmelweis had noticed that the daily event evoked a considerable reaction of worry and dread among the women in the clinic. He accepted the possibility that the disease might be psychological in origin, that perhaps the fear made the women in the first clinic more likely victims of childbed fever. He decided to test this explanation or hypothesis. He asked the priest to change his route so as not to pass through the first clinic, but this change had no effect on the mortality rate there.

Semmelweis also noticed a difference in the method of delivery used in the two sections. In the first clinic women delivered on their backs, in the second they delivered on their sides. He did not think that this difference was of any importance, but "like a drowning man clutching at a straw" he decided to test it anyway. He introduced the use of the lateral position in the first clinic but the mothers still died at the same appalling rate.

In addition to actual testing, Semmelweis rejected several other possible explanations by what we would now call "thought experiments". He had already rejected the "atmospheric-cosmic-telluric" explanation as incompatible with the evidence. Similarly, by carefully examining

other explanations about the cause of childbed fever and by comparing
them with the known evidence, he was able to reject them as well.

For example, one view was that overcrowding was responsible for
the high rate of mortality in the first clinic. Semmelweis found that
actually the second division was more crowded than the first partly due
to the desperate efforts of the women to avoid assignment to the first
clinic.

Another suggestion rejected by Semmelweis was that childbed fever
in the first clinic was caused by a sense of shame felt by the women
when they were being examined by the male medical students--the examinations were carried out by midwives in the second clinic. Semmelweis saw that if this was true, then women who delivered at home
and were examined by a male doctor should also have a high rate of
incidence. Actually the rate of puerperal fever in home deliveries was
far lower than in the hospital.

Also he rejected the idea that the difference in the incidence was
because midwives were much gentler in their examination of patients
than were the medical students. Semmelweis reasoned that the damage
age done to the vagina by the students during an examination was insig-
nificantly small compared to the natural damage inflicted by a baby
during child birth.

He continued trying to find more differences between the two clinics,
but there were more similarities than differences. The food came
from the same kitchen, the standard of cleanliness was the same in
the two clinics, the laundry was washed in the same way, ventilation
the same and so on. All the while, Semmelweis had also been carrying out autopsies on the mothers who had died of childbed fever. In
searching for a clue to the cure of this mysterious killer, Semmelweis
would arrive early in the morning and dissect the bodies of the mothers
that had died only the day before. Then he would go directly to the
first clinic with his students for the day's routine examinations. In
the Vienna hospital the autopsy rooms were adjacent to the wards.

One might think that Semmelweis's superiors would have been impressed by this young Hungarian's efforts to understand childbed fever.
But this was not the case at all. By asking uncomfortable questions
and challenging the accepted orthodoxy, Semmelweis had incurred the
disfavour of his superior, a Professor Klein. As has happened so often both before and after Semmelweis, the non-conformist, the chal-
lenger, the boat-rocker, was a threat to the establishment. In 1846
his assistantship was not renewed and Semmelweis left Vienna. Fortunately, his old post fell vacant the next year, and he was reappointed.

Back in Vienna for only a few hours, news reached him of the death of one of his colleagues. His friend, a pathologist, had just been buried after an agonizing illness which displayed the same symptoms of the many mothers Semmelweis had seen dying with childbed fever. It seems that while carrying out an autopsy, his friend had accidentally received a cut on his finger from a scalpel.

Semmelweis realized that his colleague had died from the same disease that had killed the mothers. He also hypothesized that the scalpel had introduced some "cadaveric matter" into the blood stream and caused his colleague's death. And then, for the first time, he understood why the first clinic always had a much higher rate of childbed fever. The medical students moved directly from the autopsy rooms to the first clinic to examine the women in labour. The students only superficially washed their hands with soap and water after the dissections, but the foul odour still clung to their hands for hours afterwards. Semmelweis saw that he and his students must have been unknowingly bringing death to all those mothers from the autopsy rooms. Just after childbirth, the whole womb is an open wound.

The doctors and students themselves must have been carrying death to the new mothers and the reason the second clinic had a much lower rate of childbed fever was because midwives did not perform autopsies. All the facts now fell into place, this also explained why women who had their babies at home or even in the street had a lower incidence as well.

Semmelweis made his discovery in 1847 before Pasteur, Koch or Lister had revealed to the world the importance of micro-organisms in the origin of disease. Consequently the importance of absolute cleanliness in hospitals and particularly around open wounds was not yet realized. Doctors at this time saw no more reason to keep their hands and instruments clean than do today's automobile mechanics in keeping their hands and tools clean.

Today our hospitals and doctors are so obsessed with cleanliness, that it is difficult for us to realize that this tradition is relatively new. In the 1700s doctors would not necessarily change their aprons between operations. Some deliberately kept them unwashed, feeling a certain nostalgic pride and would wear them almost like medals of past achievements. A standard joke among hospital staff was that so-and so's operating apron did not have to be hung up at the end of the day--it stood up on its own because of all the caked blood.

Semmelweis now guessed that washing with soap and water was not enough. These "particles" must be chemically destroyed. He insisted that his students thoroughly wash their hands in chlorinated lime before

the examinations. They had to wash their hands until all traces of the
cadaver aroma were gone. Semmelweis had invented the antiseptic
procedure. At first, most of his students regarded all this hand
washing as a bit silly and a waste of time.

But the results proved him right. The mortality rate in the first
clinic fell from 11% to 3%. Some of his students began to accept Semmelweis's ideas, but most of his superiors remained unconvinced.
From their point of view, there was no reason to assume that this drop
in the mortality rate was anything more than a temporary lull in the
"atmospheric telluric influence". They still regarded Semmelweis's
hand washing procedure as only a messy, useless nuisance.

Their opinion received some temporary support in October 1847
when a woman suffering from a cancer of the cervix (neck of the womb)
was admitted to the first clinic. She was put in the first bed. After
the now usual careful hand washing, the students examined first this
cancerous woman then went down the row, examining each of the remaining twelve women. To Semmelweis's horror, eleven of these
twelve women died of childbed fever. He now saw that cadavers were
not the only source of this mysterious killing substance. Semmelweis
modified his ideas by adding that childbed fever can be caused not only
by cadaveric material, but also by "putrid matter derived from living
organisms". He now required washing with chlorinated lime between
each patient.

However within a few weeks another unexpected outbreak of childbed fever occurred and Semmelweis again was able to enlarge his ideas
to explain the outbreak. This time a woman with an ulcer in her left
knee joint was admitted to the clinic. The ulcer was open, running and
gave off a foul smell. The woman's womb and vagina were not infected
and the students continued to wash their hands after each patient. Another outbreak of fever occurred and Semmelweis concluded that: "The
air of the labour room, loaded with the putrid matter, found its way
into the gaping genitals . . . and puerperal fever was the consequence."
Women with infected sores were no longer admitted to the maternity
clinic.

In 1848, the year after these two outbreaks, the mortality rate in
the first clinic fell to only 1.3%. For the first time, the student's
clinic was safer than the midwife's clinic.

Instead of convincing Professor Klein, however, all this only en-
raged him more. Semmelweis was not yet thirty, and his unorthodox
procedures had succeeded in challenging the entire body of medical
knowledge on childbirth. What is more, it seems that previously Klein

had instituted the practice of teaching students midwifery using dead
bodies--Klein's predecessor had only used a model of a woman. Semmelweis dug up figures showing that the mortality rate had risen dramatically when Klein took charge of the department. Klein was determined to get rid of this young troublemaker.

Semmelweis was granted a promotion, but without the teaching priv-
ileges normally accompanying the position. Offended, Semmelweis
left Vienna without a word even to his friends. Back in his native
Hungary, he took up an unpaid post as senior physician in a Budapest
hospital. Again he instituted his hand-washing procedures, and again
the mortality rate fell.

It was here in 1865 that he made his last addition to the understanding of childbed fever. After a sudden flare-up of fever, Semmelweis discovered that the hospital authorities had decided to save money
on the laundry bills by re-using sheets. Women were giving birth on
sheets infected with the discharges of recently deceased mothers.
Semmelweis put a stop to this by bringing an armload of these filthy,
infected sheets to the desk of the government official responsible.

In spite of his genius for discovering the causes of childbed fever,
Semmelweis had an "inborn loathing for everything that is called writing". Throughout the ten years following his initial discovery, he consistently refused to publish his findings. He had a naive belief that the
"truth will emerge". This distaste for writing proved to be his failing.
At first, a few of his friends attempted to publish his ideas for him,
but Semmelweis critized their work because they emphasized only the
early "cadaveric particles" hypothesis and did not include his later
enlarged hypotheses.

Because of his refusal to publish, his ideas remained largely unknown outside Hungary and a few of his friends in Vienna. Hungary
was the only country that officially accepted his ideas. In 1855 the Hungarian government ordered all hospital authorities to introduce Semmelweis's preventive procedures.

Finally in desperation in 1857 Semmelweis began to write his mas-
terpiece *Aetiology, Conception and Prophylaxis of Childbed Fever.*
It took him three years to write. The book was poorly written--wordy,
repetitive, rambling and egotistical. It was such a jumble that it was
largely ignored when finally published in 1861. This meant of course
that outside Hungary many mothers continued to die needlessly of childbed fever.

When Semmelweis saw that his book was not being taken seriously,
he began to write bitter, accusatory and increasingly irrational letters

to various professors of midwifery. To one professor he wrote, "So I denounce you before God and the world as a murderer". Throughout his letters he repeated over and over the phrase "The murder must stop".

By 1865 Semmelweis's behaviour had become so irrational that it was obvious that he was suffering from mental illness. Upon entering the mental hospital, the medical examination showed a cut on Semmelweis's right hand. Apparently he had received it while performing his last operation in Budapest. He died on 17 August 1865 of puerperal fever as his colleague had done some 18 years earlier. It was a slip of the scalpel that enabled Semmelweis to discover the origin of childbed fever and it was another slip of the scalpel that ended his life. The autopsy revealed extensive organic brain damage.

Five days before Semmelweis's death, a young surgeon in Scotland named Joseph Lister performed his first antiseptic operation by applying carbolic acid to the wound. Lister knew nothing of Semmelweis's work. If Semmelweis had written a clear account of his work, perhaps the name of Semmelweis would be a household word today.

Unlike Semmelweis, Lister, Pasteur and Koch published their findings and revealed to the world the nature of micro-organisms and infection and how best to control them. Like Semmelweis, they encountered initial scepticism from the scientific community, but eventually the weight of the evidence became too great for even the most sceptical of scientists to ignore. As a result, throughout the world today hospitals are clean and sanitary and doctors wash their hands so thoroughly before an operation that even Ignaz Semmelweis would have approved.

SEPARATED LEARNING

The work of Ignaz Semmelweis is just one example of the literally thousands of recorded cases of scientific discoveries. It makes a particularly good example for our purposes however because his ideas were a complete break with the currently accepted scientific explanations--he was not building on or enlarging someone else's ideas.

The reader may now be thinking to himself that Semmelweis's work was interesting and all that, but what does it have to do with the "unique type of human learning" which I promised to explain in this chapter?

Recall that in the previous chapter it was explained that there are now several accepted types of learning. At present, psychologists have universally accepted that although man's learning abilities certainly

differ in degree from those of other animals, they do not differ in kind.
Although much more extensive, man's learning ability is assumed to
be restricted to the same types of learning as used by all the other
animals.

With only two narrow exceptions all types of learning--in man as
well as other animals--are supposed to be limited to contiguous relationships. That is, learning can occur only when less than 60 seconds
separates the two connected events. When secondary learning is prevented, the rat is unable to learn that there is a connection between
the lever pressing and the dropping of the food if these two events are
separated by more than 60 seconds.

Now recall Semmelweis's problem. After being admitted to the hospital and giving birth, some mothers began to show the first symptoms
of childbed fever *two to three days* after delivery. I repeat--*two to
three days*. The time separation between an examination with hands
that smelled of cadavers and the first symptoms of puerperal fever was
two to three days. The connection that Semmelweis was searching for
and eventually found was *not* a contiguous connection. Semmelweis
succeeded in learning a non-contiguous or separated relationship.

Until now, psychologists have used the phenomena of secondary
learning to explain all learning that appeared to be separated. Remember that secondary learning occurs when a secondary event is consistently and contiguously associated with the consequence. In the Skinner
box for example, the sight of the lever was always contiguously associated with the dropping of the food for the rat. Perin (1943) showed
that when the lever was withdrawn immediately after the rat pressed it,
no learning was possible for the rat if the interval between lever pressing and dropping of the food was more than 30 seconds. This same
experiment has never been repeated with humans however.

Secondary learning cannot account for Semmelweis's discovery.
The students' examinations with foul-smelling hands were not consistently contiguous with the first onset of symptoms. If the symptoms
became noticeable within a few seconds after the infection, learning
the connection between the two events would have been simple. If the
connection had been a contiguous one, we would have learned the connection thousands of years ago, and would not have required a genius
like Semmelweis to discover it less than 150 years ago.

However, some secondary learning did occur in the story of childbed fever. Remember that the women believed that the first clinic
was more dangerous than the second. This is an example of secondary
learning. The women saw that the first clinic was more often associ-

ated (contiguously) with cases of childbed fever than was the second clinic. The women did not know the primary event or "cause" of childbed fever, but they did know that "there is something about the first clinic". This bit of secondary learning by the patients proved to be of great help to Semmelweis in his search for the primary event. Contiguous secondary events often provide useful clues to scientists looking for the separated primary event.

The connection between examinations with foul-smelling hands and the onset of childbed fever is not a contiguous connection. A contiguous connection by definition is less than 60 seconds. The first symptoms of childbed fever do not appear until two to three days after the infection. Semmelweis succeeded in learning a relationship or connection that was *not* contiguous. He was able to learn a *separated relationship*--a connection between any two or more events separated by 60 seconds or more.

Semmelweis's discovery cannot be accounted for by secondary learning, likewise his learning cannot be classified as "insight" learning. Remember that insight learning has a requirement similar to the contiguous requirement of trial and error learning. Insight learning can only occur when all elements of the problem are visually present *at the same time*.

In the previous chapter, it was discussed that if the first event had been "marked" in some way, other animals are apparently capable of learning with delays of over 60 seconds. However, marking did not occur in Semmelweis' case. Although we still do not understand exactly how other animals are capable of learning over long delays with the aid of "marking", we can get some intuitive understanding of what may be happening if we try to imagine if marking had occurred in Semmelweis' learning. Imagine that immediately after Semmelweis had examined a woman (with either clean or dirty hands), a bolt of lighting had struck just outside the hospital window. Imagine also that this occurred not just once, but with every single woman! Obviously, if this had happened, divine providence would have helped Semmelweis considerably. Marking the event would have greatly simplified the problem. The very fact that the important event had been selected and especially "marked" in some way, eliminates or at least drastically reduces the hardest part of separated learning, namely sorting through all the competing events and trying to decide which of the many possible causes is the actual cause.

Although some people may privately express some doubts about this, scientists are human. And if scientists are able to learn separated connections, then this must be considered to be human learning. Humans are capable of separated learning but other animals are not. This is the only form of learning that is exclusively human.

Of course, it has long been realized that man is the only scientific animal. But it has never been clearly understood that the learning of scientists was different *in kind* from the learning of all other animals. The type of knowledge often acquired by scientists is beyond the reach of any other animal--even in part. All other animals are limited to contiguous learning, man is capable of both contiguous learning and separated learning.

To put this in the form of a precise statement, the proposal of unique human learning is that: Humans are the only animals that are able to learn by trial and error or experimentation that a relationship or connection exists between two or more events that are separated by time (more than 60 seconds). In other words, only humans are capable of separated learning. This statement assumes, of course, that secondary learning has been prevented and that the events concerned have not been "marked" in some way.

Actually contiguous learning is, in all likelihood, limited to much less than 60 seconds, probably even less than 30 seconds. Whether contiguous learning is limited to 5 seconds or 21 seconds or 34 seconds or 44 seconds is not too important here. This detail will have to be settled experimentally. The important point is that it is certainly less than 60 seconds.

Critics may argue that learning with a separation of less than 60 seconds as opposed to a separation of 2 to 3 days is only a difference of degree rather than a difference of kind. However, this is not the meaning of the word "kind" used here, or the meaning given to this word by scientists working in this area in the past.

Remember that Chapters 1 and 2 reviewed the various abilities which have been proposed in the past to mark man's intelligence as different in kind from that of the other animals. From reviewing the history of this "kind versus degree" dispute, it is clear that the men who have proposed a difference of kind between animal and human intelligence meant by this that humans had an objective, identifiable abiiity which other animals could *never* master. They proposed that there was a barrier which other animals could never cross. Humans were capable of doing something which no other animal was ever capable of doing.

For example, until the recent experiments (discussed in Chapter 2)

in which chimps were successfully taught several types of symbolic
languages, several scientists proposed that human intelligence was different in kind because we were the only animals capable of using symbolic language. By this it was meant that although humans were capable of using symbols for communication, other animals could *never* do
so. However, when Washoe and other chimps were able to learn and
use even a few symbols, this proposed difference of kind was disproved.
This proposed difference of kind has now been shown to be only a difference of degree. By using even a few symbols, chimps have revealed
that they could cross this proposed barrier. *

Scientists who proposed that humans were the only animals able to
use symbols, never claimed that other animals were not capable of any
type of communication--it was long accepted that they were capable of
a certain amount of communication by using gestures and genetic display signals. The claim of a difference of kind was that other animals
were *never* capable of a *particular type of communication*, namely symbolic language.

Proponents who have argued that human intelligence is different in
kind have defined quantitative differences in this way, that we have a
demonstrable ability which no other animal is *ever* able to perform
(such as tool-making or using symbolic language). Once another species has been able to perform one of these activities *even on a very
limited or simple level*, then it is no longer possible to claim that this
is a difference of kind.

This is exactly the same meaning of the word "kind" used here. I
have proposed that humans have a unique ability which other animals
are *never* able to perform. Obviously all animals are capable of some
types of learning, but humans are capable of a specific type of learning
which other animals can never master. Other animals are capable of
contiguous learning (less than 60 seconds) between any two or more
events but beyond this barrier they can never pass. Only humans are
capable of "separated" learning (more than 60 seconds).

All animals learn about their environment in order to increase their
chances of survival. Animals in the wild learn to avoid dangerous locations or predatory animals. They learn from experience where to
go to find food or water. Similarly when put in a Skinner box, animals

---

*However, as also discussed in Chapter 2, I have proposed that the
actual difference of kind between humans and other animals in this area
is not the *using of* symbols, but rather the *creation* of them (symbolizing).

are able to learn that they can obtain food by pressing a lever. Learning is a method whereby individual animals are able to change their behaviour, often helping them increase their chances of survival. Individual animals are able to quickly adapt their behaviour to suit the given environment.

Like all other animals, humans also are constantly trying to learn about our environment in order to help us survive. We try to learn how to prevent events which endanger our survival such as disease and starvation and try to learn how to increase events that are favourable to our survival. Like the rat in the Skinner box, we attempt to survive by learning about our environment. For scientific purposes, a rat's learning may be limited to the Skinner box, but our learning is not limited to a psychologists's laboratory. The whole universe is our Skinner box. Like the rat, we attempt to learn about our environment and then use this knowledge when possible to ensure our survival.

When the rat observes the food pellet drop into the tray for the first time, this is a mysterious, unexplained event for him. He does not know the "cause" of this fortunate event. It just happened. He does not yet know that pressing the lever causes this happy event to occur. There is no observable string or rope connecting the lever with the food tray. He is not able to see the mechanism which allows a pellet to fall when the lever is pressed. There is no visible connection between the two.

Similarly, when Semmelweis first started studying childbed fever, this was a mysterious event to him. Some women began to show the telltale symptoms, others did not. These symptoms just mysteriously appeared with no apparent connection with anything else. There was no observable connection between an examination with foul smelling hands and the onset of symptoms. There was no visible connection between these two events. Like the rat, he did not observe the mechanism (which we now call the germ) which connected these two events.

Although we are like the rat and all other animals in trying to learn about our environment, there is one very important difference. All other animals are able to learn connections between events only if these connections have an interval of less than 60 seconds. All other animals are limited to contiguous learning. Man however, is able to learn connections between events that are separated by not only seconds, but also by hours, days, weeks and even years.

After being placed in a Skinner box, a rat will sooner or later hit the lever. If the lever immediately withdraws into the wall (to prevent secondary learning) and then the rat is left undisturbed in the experi-

mental cage for say five minutes before the food drops, the rat will
never be able to learn the connection. This sequence can be repeated
indefinitely, but *learning will never occur.*

On the other hand, if a similar situation is set up with human subjects, learning can occur. The simplest experiment would be to repeat as closely as possible Perin's experiment in which he attempted
to eliminate secondary learning in rats. Details of his experiment can
be found in his paper. This experiment could be set up in the following
manner: having arrived for an appointment with a doctor or a teacher,
the person is seated alone in a small, plain waiting room and told it
will be a few minutes before he can be seen. On the wall is a lever or
button, nearby is also a small square opening in the wall. Of course,
no instructions or explanation would be given to the person, as a rat
never has the benefit of instructions. After the lever is pressed, the
lever immediately withdraws into the wall, and five minutes later a
chocolate bar appears in the opening in the wall.

Of course, this situation would have to be repeated on several different days before learning could occur and be confirmed. Even if
repeated many times, some people would never succeed in learning
the connection, indeed some people would never even press the lever.
But at least a few people would be able to learn this separated connection.

In all likelihood, people differ widely in their separated learning
ability. Although hundreds of doctors were concerned with childbed
fever, only Semmelweis succeeded in learning the connection, and it
was eighteen years after Semmelweis's original discovery that another
doctor succeeded in rediscovering the nature of infection.

Also it seems quite likely that experimental results will show that
the longer the interval between events, the more difficult learning will
be. Learning a connection with an interval of only 5 minutes will be
much easier than with an interval of 5 hours and an interval of 5 hours
will be easier than an interval of 5 days and so on.

This experiment would be relatively simple to perform, and I invite
psychologists to test it. It would also be possible to repeat Kohler's
insight experiments with humans (see p. 79-80). Kohler found that
chimps were unable to achieve an insight solution unless all the necessary parts of the problem were directly visible *at the same time.* Unless the chimp was able to perceive the stick and the food in the same
glance, the animal was unable to use the implement.

Separated learning has occurred too many times in human history
not to occur again in a psychologist's laboratory. Actually, separated

learning has probably occurred in psychologist's and other scientist's laboratories more often than anywhere else. When separated learning occurs, we usually call this a "scientific discovery". In the above proposed experiment, humans will be able to *learn* (that is, change their behaviour towards the lever or stick), whereas experiments have already shown that other animals in the same situation are unable to do so. Obviously Semmelweis changed his behaviour towards "cadaveric matter" after his learning experience.

"Higher" human learning was viewed as a "cognitive" type of learning. However, both the main antecedents of modern cognitive learning theory, the gestalt insight theory (see Kohler, 1927 who conducted his experiments on chimpanzees) and the sign learning theory of Tolman (1932) who mainly worked with rats, agree that learning in these animals must be contiguous. No cognitive theorist working with humans has proposed that human learning might be an exception to the principle of contiguity. No cognitive psychologist has ever bothered to repeat Perin's or Kohler's experiments with humans.

The reason psychologists have never been fully aware of this type of learning is that it occurred in themselves, rather than in the subjects they were studying. They couldn't see it because they were not observing themselves. They ceased being behaviourists when it came to their own learning. They viewed other animals' learning from the "outside" (behaviour) but still viewed human learning from the "inside" (cognition). Because psychologists looked at human learning from the "inside" and conducted different types of experiments on humans as a result of this different outlook, they never realized that we were exempt from the principle of contiguity.

My proposal is that if only for consistency's sake, we should consider human learning in the same terms which we use when studying the learning of all the other animals. We should consider human learning from the "outside" (in terms of behaviour). As psychologists we should consider science from a behaviouristic point of view. As scientists we should consider science from a truly scientific point of view.

The scientific study of man and also of learning is mainly carried out by psychologists. When studying human subjects, psychologists often observe their subjects from an adjacent room through a one-way mirror. Perhaps at least occasionally, it would have been helpful if psychologists had also installed a regular mirror in their observing room. Then they could have observed their own behaviour as well, and perhaps it might have occurred to them that they should be consid-

ering human learning in the same terms in which they study the learning in all other animals.

Perin published his study of contiguous learning in rats in 1943,
which is now almost forty years ago. If Perin or anyone else had repeated the same experiment with human subjects, the fact that humans
are unique in their learning ability would have been discovered long
ago. The fact that no one has repeated Perin's experiment with humans, reveals one of the major faults in attempting to advance scientific knowledge with only the "blind experimentation" which I referred
to in the Preface.

A scientist conducting experiments "blindly" may occasionally stum-
ble across an important discovery, but when he does, it is often pure
luck. Often he may come perilously close to a major discovery, only
to turn away and begin searching somewhere else. But a near miss is
still a miss. Perin came very close to discovering separated learning,
but he did not realize the importance of repeating this experiment with
human subjects. Often blind experimentation is necessary especially
in a new science, for example Semmelweis had no idea what he was
looking for and was proceeding by blind experimentation. Luck played
its part in his discovery. It is questionable whether he would have
succeeded without his colleague's death.

In the first stages of a young science blind experimentation is necessary, but theoretical formulations are also necessary for a maturing
science. Broad theoretical formulations bring the experimental evidence into a framework and point out important areas for the experimental scientists to study. Instead of conducting their experiments
almost at random, a theory enables the experimental scientists to direct their efforts to particular phenomena. A good theory creates order and simplicity out of what was previously only a chaotic mountain
of unrelated facts. Great advances in science are almost always advances in theory. Copernicus, Newton, Darwin and Einstein were all
primarily theorists.

Many of today's learning psychologists maintain that broad theories
are not necessary in psychology. They often try to make a virtue out
of their exclusively blind experimental approach. If behaviouristic
learning psychologists did not have such an abhorrence of theory building and anything resembling philosophical speculation, perhaps separated learning would have been discovered long ago.

Humans then, have not just one unique ability, but two. We are not
only capable of symbolic language, but also of separated learning. Although language is the most obvious difference between humans and the

other animals and this difference has been recognized by men for a
very long time, it is only in this century that it has been possible to
ferret out the exact difference between our form of language and the
communications used by other animals. Only recently have we been
able to classify and define exactly what makes our symbolic language
different *in kind* from the types of communication used by all other animals. Similarly, men have long realized that man's learning ability
was far more extensive than the learning ability of any other animal.
It has also long been realized that humans are the only scientific animals. But only now has it been proposed that this scientific or separated learning ability is a completely different type of learning and
that this new type of learning is the exclusive domain of the Homo
sapiens species.

Chapter 5

Hypotheses, Theories & Foresight

Separated or scientific learning in human societies occurred long before Semmelweis and has continued right up to the present day. As a matter of fact, at least in Western societies, the rate of separated learning has appeared to speed up dramatically in the past few centuries. Semmelweis's work on childbed fever was only one example out of thousands of known cases of separated learning by humans.

Another equally good example, also from medical science, occurred in 1900 when Walter Reed discovered the connection between mosquito bites and yellow fever. Working with American troops stationed in Cuba, Reed conducted a series of carefully controlled experiments showing that the mosquito was responsible for the spread of yellow fever.

Before Reed, the most commonly accepted hypothesis concerning the onset of yellow fever was that of "fomites". Fomites were supposed to be germs that were given off by an infected patient and were then able to grow in nearby articles of clothing, bedding and furniture. Yellow fever was believed to spread when these fomites were transported in baggage or merchandise to distant locations.

Hence, health authorities spent a great deal of time "disinfecting" clothing and bedding of yellow fever patients. Expensive quarantine regulations were also enforced in an attempt to prevent these fomites from entering a port through the mail, baggage, freight, ships and so on. Reed showed that all these regulations were unnecessary by demonstrating that yellow fever was caused by the mosquito and not fomites.

A soldier who had been confined to a military jail for over a month contracted yellow fever and died. Eight other prisoners also occupied this same cell and one even slept in the same bunk recently vacated by the sick man, yet none of these prisoners contracted yellow fever. Reed saw that if the fomite hypothesis was correct, some of the other prisoners should also have come down with yellow fever, since they were in close and continuous proximity with the sick man and his bedding. Reed proposed an alternative hypothesis that "perhaps, some in-

sect capable of conveying the infection such as the mosquito, had entered through the cell window, bitten this particular prisoner, and then
passed out again. This however, was only a supposition" (quoted in
Kelly, 1907, p. 125).

A careful investigation was then conducted to test this hypothesis.
Volunteer soldiers were then infected by either (a) bites from mosquitoes which had previously bitten a yellow fever patient or else (b) close
contact with sheets, blankets and so forth which had been contaminated
with discharges of patients with yellow fever. Dr. Reed comments
that many of these articles "had been purposely soiled with a liberal
quantity of black vomit, urine, and fecal matter".

Although the volunteers who slept in the "Infected Clothing Building"
near these supposedly unhealthy articles probably had a much harder
time sleeping than did their friends sleeping in the "Infected Mosquito
Building", it turned out that only those soldiers that had been bitten by
infected mosquitoes contracted yellow fever.

Dr. Reed had shown that yellow fever could only be contracted from
mosquitoes and never from "fomites". Like Semmelweis, Reed became convinced that the prevailing scientific hypothesis could not account for the evidence he had observed. Having rejected the generally
accepted hypothesis, he postulated another which was compatible with
the evidence he had so far observed. He then carefully tested both
hypotheses.

Reed succeeded in discovering a separated relationship. He later
found that the period of incubation for yellow fever was about three to
six days. That is, it was three to six days after the mosquito bite before the patient began to feel the first chills of yellow fever. As with
childbed fever, if the relationship had been a contiguous one, learning
would have been easy. If the patient had begun to feel chills within a
few seconds after the mosquito bite, humans (as well as other animals)
could have learned the connection long ago.

Neither Semmelweis nor Reed were aware of the mechanism that
created the relationship they discovered. Although micro-organisms
were first seen in a microscope more than a hundred years before
Semmelweis was born, the causal role of micro-organisms in the origin of disease was not yet known when Semmelweis was studying childbed fever. He did not know *why* examining pregnant women with hands
that smelled of cadavers caused childbed fever. He only knew that
there was a connection between the two. By the time Reed was doing
his work on yellow fever however, the function of germs in the role of
disease was well known. It was widely assumed before Reed that a

particular germ caused yellow fever. But the specific germ causing
yellow fever was not found until *after* Reed had proved that there was
a connection between mosquito bites and yellow fever. Often scientists
succeed in first discovering a separated relationship and only later uncover and name the mechanism which produced the ralationship.

The two examples of separated learning cited so far have both been
relationships with intervals of only a few days, but humans are by no
means limited to learning only connections separated by days. We are
able to learn relationships separated by weeks, months and even years.
For example, it was Benjamin Franklin in 1784 who first hypothesized
that eruptions of volcanoes could have an effect on the weather in the
following year or two. The volcano ejects tons of fine ash into the stra-
tosphere during eruption and this pollution is purported to reduce the
amount of sunlight reaching the earth, as a result reducing surface
temperature as much as two years later. This hypothesis is now accepted by many scientists and if valid, could have important consequences for modern man because of the tons of man-made pollution we
pump into the atmosphere every year.

An example with an even longer interval is revealed in the history
of our understanding of syphilis. Although doctors first recognized
syphilis in 1495, it was only in the 1820s that it was first suspected
that there was a connection between a syphilitic infection in early adult-
hood and the onset of general paralysis and insanity in middle age. We
now know that between the second and third stages of syphilis there is
a latent period with no symptoms whatsoever. This latent period sometimes lasts up to 30 years or more. After the latent period, the third
or tertiary stage often sets in which is usually highly destructive to the
heart, brain and spinal cord.

The learning of separated relationships is not confined to the past,
of course scientists are still attempting to discover more separated
connections today. For example, Russian scientists have recently discovered an apparently important separated relationship concerning
earthquakes. In searching for a way to predict earthquakes, the scientists began studying the area near known faults in central Asia. They
discovered that before an earthquake the "seismic velocity" of underground rocks (the speed at which the rocks transmit sound waves) near
the fault was temporarily reduced by something like 10%. When this
seismic velocity returned to normal, an earthquake occurred soon
afterwards.

It seems that the interval between the drop in seismic velocity and
the earthquake varies with the magnitude of the earthquake. A very

large earthquake apparently gives more than ten years warning, but a very small earthquake gives only a few days notice. But with all earthquakes, the seismic velocity returns to normal just before the quake.

This proposal that a connection exists between the seismic velocity of rocks and earthquakes has yet to be carefully tested and confirmed. Much careful observation and experimentation will have to be conducted before scientists will accept that there is indeed a relationship here. But whether or not this hypothesis turns out to be true, even in part, is not of central importance here. In either case, this is a good present day example of how science advances by proposing connections between separated events.

Previous examples used have primarily been concerned with the separated connection between *two* events. Note that in this example, there are *three* events. The first event is the drop in the seismic velocity, the second event is the return of the velocity to normal and the third event is the earthquake. The scientists have proposed that a certain relationship exists between these three separated events.

Concerning the explanation for this relationship, scientists now believe that the temporary change in the sound velocity of the rocks is caused by the strain these rocks undergo before an earthquake. But this explanation was only put forward in 1973 *after* the original relationship had been proposed. As has happened so often in science, the hypothesis of a separated relationship between events came first, and only afterward was the mechanism explaining this relationship identified.

The search for separated relationships continues in all areas of modern science. For example, the extensive and expensive search for the causes of cancer, arteriosclerosis, and the many forms of mental illness can all be viewed as a search for separated relationships.

THE SEPARATED HYPOTHESIS

Given that humans do indeed have a unique learning ability--that we are the only animals able to learn separated relationships, we now may ask ourselves how is man able to accomplish this feat? What is the mechanism which allows man, and only man, to learn separated connections?

When other animals are attempting to learn about their environment, they often do not make only meaningless, confused random and uncoordinated responses. Rather, as Krechevsky (1932) found, animals often adopt systematic modes of solution to learning problems. In-

stead of making random and unrelated responses, animals often go
about learning in an orderly, systematic manner. These early systematic attempts or false solutions used by animals in learning were
named "hypotheses" by Krechevsky. Krechevsky was somewhat reluctant to use this word in relation to animal learning however, and
consequently always enclosed the word in quotation marks.

But if all other animals are limited to contiguous learning, it follows that they are only able to formulate "hypotheses" between events
that are contiguous. Since no learning occurs in other animals if the
events are separated by more than 60 seconds, it follows that they
must be unable to formulate hypotheses between separated events. Put
in gestalt terminology, other animals are able to "hypothesize" only
if all the elements of the problem are visually present *at the same
time.*

In attempting to explain why other animals' learning is restricted
to contiguous relationships, Spence (1956) and Revusky (1971) proposed
that with longer and longer intervals, the animal experiences more and
more events which compete with the primary event. With a long interval, the animal is unable to hypothesize which of the many events
in his past caused his reward or punishment. This explanation has received some experimental support. When rats were kept in close confinement during the interval, learning was significantly higher than
when the rats were not kept in close confinement. Apparently, the
close confinement prevented the animal from experiencing as many
competing events.

This problem of numerous possible previous events is exactly the
problem a scientist faces in attempting to discover a separated relationship. In Semmelweis's search for the causes of childbed fever, he
was confronted with many possible causal events. Perhaps it was overcrowding, or the priest passing through the ward, or perhaps the method of delivery or a feeling of guilt in the women and so on. There
were many possible events.

Semmelweis proceeded by making various proposals or hypotheses
about which past event might be the primary event. He made several
hypotheses about which event might be related to the onset of childbed
fever. After each hypothesis, Semmelweis would test this proposal by
either actual experimentation or else a thought experiment. After one
hypothesis proved not to be supported by the evidence, he tried another and then another and so on until he found one that was supported by
his testing.

So although other animals are able to "hypothesize" between contiguous events, they evidently are not able to do so when the events are separated by time. However, humans are capable of formulating hypotheses between events separated by time. Semmelweis proposed a hypothesis between two events that were separated by two to three days. Other animals are restricted to contiguous "hypotheses", but only humans are capable of formulating separated hypotheses. Only humans are able to formulate true hypotheses (without quotation marks).

We see then, that the use of the true hypotheses is essential in any search for a separated relationship. A true or separated *hypothesis* is any statement that proposes a relationship or connection between two or more more separated events. Without the hypothesis, the search for separated relationships and much of what we call science would not be possible. It is the hypothesis which proposes a link between one or more of the many possible competing previous events and the consequence. The hypothesis brings together (if only in the mind of the scientist) two or more events which are actually separated from each other in time.

After a hypothesis has been formulated, the scientist proceeds to test it. Since the time of Semmelweis, science has made enormous strides in the perfection of the methods of testing hypotheses. The use of several types of control groups, the use of statistical analysis of data and so on have been studied and formalized only recently. The sophisticated techniques of experimentation have helped scientists greatly in their ability to test hypotheses. These methods have already been adequately described in many scientific textbooks and therefore I will not attempt to explain any of these methods again here.

As important as testing and experimentation are to science, they are no more important than is the process of hypothesis formulation, however. As Popper (1963) has pointed out, without a hypothesis, an experiment is not possible. The first step in any scientific inquiry involves deciding which events might be important for study. It is not possible to run an experiment without first deciding which events you are going to experiment with. It is not possible to experiment or even observe without first proposing which few events (out of the millions of events in the universe) might be important for your purposes.

This is not to say that the process of hypothesis formulation or hypothesizing has always been fully and consciously recognized. Even to this day a few scientists still believe that they can go into a laboratory and collect "raw" experimental data without the aid of hypotheses or theories. But these scientists are deceiving themselves. Even these scientists must form at least a tentative, hazy, semi-conscious hunch

that "X might be related to Y" before they could perform an experiment testing whether X was indeed related to Y. Often hypothesizing
is not a fully conscious process, but it is nevertheless essential before
any testing or even observation is possible.

This ability to hypothesize then is our second unique ability. Man
has not one unique ability, but two. As was explained in Chapter 2,
we are the only animals able to symbolize, that is to create symbols.
We now see that humans also have a second unique ability--the ability
to hypothesize--to propose that a relationship or connection exists between separated events. Only humans can symbolize and hypothesize.

Usually the term hypothesis is restricted to the field of science. In
common usage, a hypothesis refers only to statements made by scientists. Note however that I have defined a hypothesis as *any* statement
which proposes a connection between separated events. This is a considerably expanded definition. Not only does this expanded definition
include proposals made by scientists, but also statements made by
housewives, businessmen, coal miners, children, preachers and even
witch-doctors in primitive tribes. Hypotheses are not only scientific
proposals about nature, but also superstitious statements, old wives'
tales, and magical proposals. Properly speaking, however, religious
statements cannot be classified as hypotheses; this distinction will be
explained in Chapter 8. Most if not all people form hypotheses in their
everyday lives. The ability to hypothesize is not limited to scientists
or even to educated people.

If a farmer says "I can always tell when it is going to rain, the day
before it rains, my legs always begin to ache". This statement is a
hypothesis. It proposes that there is a connection between his aching
legs and rain the next day. In the early 1970s when the United States
was making moon landings every few months, there also was some
particularly turbulent weather over parts of America. Some people
proposed that the moon landings had in some way caused this bad
weather and demanded that the moon flights be halted. This was a hypothesis proposing a connection between moon flights and the weather
on earth.

A hypothesis is only a *proposal* about how things are, a guess or a
supposition if you like. When tested, many if not most new hypotheses
are not supported by experimental results. Until they are tested, one
hypothesis must be regarded as as good as any other. It is not possible to categorize a hypothesis as good or bad until an experiment has
been performed to test it (either an actual experiment or else a "thought
experiment" using already available evidence). A street sweeper is

just as able to propose hypotheses as is the highly trained and educated
scientist.

But surely, some people will say, a hypothesis formulated by a scientist is more likely to be supported by experimentation than is a hypothesis formulated by a street sweeper. This *may* be the case, but I
would not care to defend this assertion too strongly. Scientists have
often proposed hypotheses that in retrospect appear to be as ridiculous
as any superstitious old wives' tale. For example, Erasmus Darwin
is said to have believed that during sexual intercourse, if the man
thought about females during the sex act, then the resulting pregnancy
would produce a baby girl. Needless to say, this hypothesis has not
received any empirical support.

Also, although scientists usually turn up their noses at any "superstitious" beliefs of uneducated people, it has turned out that some of
these "superstitious" hypotheses have eventually received empirical
support. As a doctor in England in 1788 Edward Jenner heard the old
wives' tale that "people who have had cow-pox do not get small pox".
This old wives' tale turned out to be true and because he did not dismiss this "superstition" out of hand, Jenner was eventually able to invent vaccination. Also remember that most of the doctors that worked
with Semmelweis in the Vienna General Hospital dismissed the women's
fear of the first clinic as unfounded.

However, I would defend the assertion that a hypothesis formulated
by a scientist is much more likely to fit within the currently existing
scientific theories than is a hypothesis formulated by a street sweeper.
All that long and expensive training does accomplish something. A
scientist trained in one scientific theoretical framework is much more
likely to propose hypotheses that fit within that framework than is an
untrained person.

Broad scientific theories provide a comfortable framework for scientists to work in. Theories direct the scientists' attention to certain
important areas and concentrate their efforts on particular problems.
Theories are necessary and extremely useful in the course of scientific advancement, but they are by no means an unmitigated blessing.
In addition to their many positive benefits, they also have some disadvantages. The main disadvantage is that they tend to narrow the
scope and originality of most scientists, somewhat like putting blinkers
on a horse. Once a scientist has accepted a particular scientific theory,
he is much more likely to notice only those facts which fit in with that
theory. He is likely to play down the importance of or even ignore
those events which refuse to fit within the framework. This process

has been explained and discussed by Thomas S. Kuhn (1970).

Scientific breakthroughs occur when one scientist refuses to ignore these "difficult" facts. Semmelweis was such a man. He rejected the currently accepted scientific explanation because he realized that it could not account for the evidence he had observed. He then began to propose hypotheses that did not fit into any theory. Most scientists apparently find it extremely difficult to break out of the framework imposed by the presently accepted theory.

This narrowing function of accepted theories exists not only for the professional scientist, but also for other people as well. In all cultures there exists a general framework of beliefs concerning what type of events cause other events. In primitive societies many of these broad cultural theories are usually what we call superstitious or magical beliefs. In our Western societies, generally speaking, scientific theories have replaced these earlier beliefs even among the general population. Like the professional scientist, people in all cultures tend to propose hypotheses that fit within their accepted theories or explanations, whether these theories be of a superstitious, magical, religious or scientific nature. The relationship between magic, religion and science will be examined in more detail in Chapter 8.

For example, if a man in a Western society wakes up with a bad cold, he may well begin asking himself what might have caused his cold. He may well propose several possibilities to himself. Did he get chilled at any time the previous day? Did he get his feet wet walking home from work? Has he been getting enough sleep? and so on. He is of course proposing various hypotheses to explain separated events. It is important to note that all these hypotheses fit within our currently accepted medical explanation concerning which events are likely to cause colds. It is highly unlikely that a man in our society in this situation would ask himself what colour shirt he wore yesterday or if he took the bus to work yesterday. Once a particular explanation has been accepted by a person, this tends to limit the hypotheses he may propose about the relevant events. A person in a primitive society who woke up with a bad cold might well ask himself entirely different questions from the ones we would probably ask ourselves. For example, a man from a primitive culture might well ask himself if he had any enemies that might be playing sorcery on him or if he had broken any taboos recently and so forth.

Separated learning is always easier if the correct hypothesis fits within the currently accepted cultural theories or explanations. For example, in the previous chapter I proposed a test to demonstrate separated learning in humans. The relationship to be learned was between pressing a lever and the dropping of a chocolate bar. Certainly this is a commonly accepted connection in our society. Vending machines work on this principle and many people are generally familar with how a Skinner box works. Forming a hypothesis between these two separated events should come very easily to most people in our society. On the other hand, if the experiment was set up so that the chocolate bar appeared several minutes after the person has coughed or yawned or tapped the table with his hand or some such behaviour, separated learning will be much more difficult. Or in a society where people were not familar with vending machines or Skinner boxes, learning would be much more difficult for these people than it would be for us in the original lever pressing experiment.

New hypotheses can be divided according to those that fit within the currently accepted scientific theories and those hypotheses that do not. But a hypothesis that fits into the currently accepted theory will not necessarily be supported by experimental testing, and those hypotheses lying outside the accepted theory may turn out to be valid.

All societies whether "primitive" or advanced have proposed numerous hypotheses about the nature of the universe and relationships between events in the universe. No society can be said to be seriously lacking hypotheses explaining why important events occur. A man from a primitive tribe will almost always have an explanation for an event that is important to him. Ask him why there is lightning or why the crops didn't grow well last season and almost without exception he will give you at least one explanation, if not several. True, most of his explanations are superstitious or magical or religious in nature, but they are nevertheless explanations. The ability to hypothesize appears to be universal in every culture.

What distinguishes our separated learning ability from the learning ability of primitive peoples is not in the nature or quality of the hypotheses we propose, but rather in what we do with our hypotheses once we have them. The idea of accepting a hypothesis only after we have tested it is what separates science from all other endeavours. The concept of admitting our ignorance about what causes a certain phenomenon to occur (in other words of not accepting any hypothesis) is only a very recently accepted conception. The notion of not accepting a hypothesis until it has been tested is restricted to societies where science is a major endeavour.

Scientists have long proposed that hypotheses should be formulated so that they are testable, that is so that they can either be supported

or not supported by empirical evidence. Many hypotheses are formulated in such a way that it is not possible to test them. If someone
proposed that rabbits were able to cause rain by looking at the clouds
and emitting "rain waves" from their eyes, this would be a hypothesis.
But unless scientists were able to detect these mysterious "rain
waves" with some sort of instrument or test this hypothesis in some
other way, this would not be a testable hypothesis.

From the point of view of science, testable hypotheses are much
superior to non-testable hypotheses. But this is only a scientific value
judgement. Only testable hypotheses are capable of becoming scientific hypotheses. But there is no reason to assume that a non-testable
hypothesis is necessarily false. For all we know, rabbits might really
emit "rain waves", but until we devise a method of testing for these
waves, this is not a testable hypothesis, and hence cannot be supported
by empirical evidence.

It has happened that hypotheses that were formerly non-testable,
become testable due to the invention of a new instrument such as a telescope, microscope or geiger counter. Only a few decades ago, the
hypothesis that "the seismic velocity of rocks changes just before an
earthquake" would have been a non-testable hypothesis. Yesterday's
non-testable hypothesis might become today's testable hypothesis and
then tomorrow's scientific hypothesis. Some examples of non-testable
hypotheses that have become scientific hypotheses are atomism and
terrestrial motion.

Although science prefers testable hypotheses, it is not possible to
claim that a new testable hypothesis is any more likely to be true than
is a new non-testable hypothesis. All hypotheses are, after all, only
guesses about what unseen relationships might exist in nature. A non-
testable hypothesis is not in any ultimate sense inferior to a testable
hypothesis, both *might* be true. A testable hypothesis that has not yet
been tested, and a non-testable hypothesis must be regarded as having
an equal chance of being true. Although many of the hypotheses proposed by primitive peoples are non-testable, we cannot claim that
these hypotheses are in any ultimate sense inferior to testable hypotheses.

The basic ability to propose hypotheses is universal in all human
cultures. Just as no culture can be said to have an inferior or primitive art or language, it is also true that no society is lacking in the
ability to hypothesize, to propose that certain separated events are
related to each other.

What differentiates modern man from primitive man is not the abil-

ity to hypothesize, but rather the notion that a hypothesis must be carefully tested before it should be assumed to be valid. Science can thus
be viewed as a "learning set", a slowly acquired knowledge of how best
to learn separated relationships. Only painfully slowly have men gradually and grudgingly accepted the notion that a hypothesis should not be
assumed to be true just because it is supported by the political or religious leaders or because it fits in nicely with our preconceptions.

Humans have gradually learned how best to learn separated relationships. Although the ability to propose hypotheses is universal in
all cultures, the clear knowledge of the best criterion for accepting or
rejecting a hypothesis is only a relatively recent aquisition. Only very
slowly through the centuries have humans found that hypotheses that
are accepted only after they have been carefully tested prove to be
more useful than those hypotheses that are accepted for any other reason. The very foundation of science itself then is a hypothesis about
which hypotheses will prove to be most useful and functional to the
survival of mankind.

THEORIES

Although theories have been mentioned in passing, they have not as
yet been discussed or defined on their own. Exactly what is a theory
and how does it differ from a hypothesis?

It is not possible to draw an exact line of demarcation between hypotheses and theories. They are similar creatures. Both hypotheses
and theories propose relationships between separated events. On the
simplest level, a theory is often only a large scale or overgrown hypothesis. Generally speaking, hypotheses propose relationships between a small number of separated events and theories propose relationships between a large number of separated events. The ability to
hypothesize therefore includes not only the ability to formulate hypotheses, but also the ability to formulate theories.

The more events a theory attempts to relate to each other, the
broader or more general the theory is said to be. Theories range
from narrow or low level theories dealing with only a limited set of
phenomena to the large, general theories that attempt to relate hundreds, thousands or even millions of diverse separated events or phenomena. Newton's theory of universal gravitation is a good example
of a general theory. Newton proposed that all bodies in the universe
attract each other in a particular way--namely that the force of attraction varies inversely with the square of the distance between the bodies.
The fact that Newton proposed that *all* bodies in the universe are re-

lated to each other in a particular way shows how general his theory
was. This theory not only proposed relationships between a falling
apple and the earth, but also between the sun, the moon, the planets
and even the most distant stars. His universal theory of gravitation
proposed a relationship between literally millions of events.

The theory of evolution proposed by Charles Darwin is also a general theory. Darwin proposed that the origin, survival or extinction
of all species of living things--animal and plant--can be viewed as a
"survival of the fittest". Thus, the existence of not only thousands of
different species of plants and animals now living, but also the many
fossils of long extinct species were related to each other.

This book is in itself a general theory--a proposal that many of the
differences of behaviour between humans and all the other animals can
be related and explained by man's unique mental capacities.

Even in large or general theories, science never attempts to find
the ultimate causes or truths of the universe. Science only attempts
to describe relationships between observables. When science "explains" something, it simply proposes that this thing is related in a
certain way to other phenomena. For example, Newton never attempted to explain why all bodies attract each other, in other words
why gravity exists. He only proposed that bodies attract each other in
a particular way. The ultimate questions such as why gravity exists,
or why matter exists or why regular relationships appear to exist between objects in the universe is forever beyond the range of science.
Even after Newton and Einstein, modern scientists are still just as
totally ignorant about why an apple falls to the ground as were men
thousands of years ago.

As with hypotheses, no culture can be said to have "primitive" theories. All cultures and all peoples are capable of formulating hypotheses and quite complex theories. A new hypothesis or theory proposed by a witch-doctor cannot be said to be primitive or inferior to a
new hypothesis or theory formulated by a modern scientist. Only after
empirical evidence is found to either support or refute a hypothesis or
theory can they be classified better or worse than any other proposal.

In the history of scientific investigation, a new field of study often
begins with "blind" experimentation and eventually yields several scattered hypotheses that have been empirically supported. Often these
hypotheses do not appear to be related to each other. Again we can
use Semmelweis's work as an example. Semmelweis's first supported
hypothesis was that there was a separated relationship between "cadaveric matter"·placed in an open wound and childbed fever. This hypoth-

esis will be diagrammed as follows:
1. "Cadaveric matter" $\xrightarrow{\text{(2-3 days)}}$ childbed fever.
    A later outbreak of fever required Semmelweis to postulate another hypothesis:
2. "Putrid matter from living organisms" $\xrightarrow{\text{(2-3 days)}}$ childbed fever.
    And yet another outbreak required him to make yet another hypothesis that it was not necessary for this "putrid matter" to be put in direct contact with the open wound, but that it could pass through the air.
3. "Air . . . loaded with putrid matter" $\xrightarrow{\text{(2-3 days)}}$ childbed fever.
    Each of these three separate hypotheses which Semmelweis had proposed had received some empirical support. But since the role of germs had not yet been recognized in the spread of infection, to Semmelweis these three hypotheses must have appeared to be separate and almost unrelated to each other. Semmelweis had no real central concept to explain to him why each of these three hypotheses had been empirically supported. Semmelweis had no idea why each of these three hypotheses (out of the many he had formulated) had been supported. For all he knew, the next hypothesis that would be supported would be that any strong smell near open wounds could cause childbed fever.
    Of course, we can see now that each of his three supported hypotheses is "explained" by modern germ theory. Germ theory proposes that all infections are caused by micro-organisms. Although Semmelweis did not live to see it, other men (in particular Pasteur, Koch and Metchnifoff) were able to formulate germ theory out of the results of many supported hypotheses.
    As any young science develops, it gradually accumulates more and more hypotheses that have received empirical support. Inevitably, as these unrelated hypotheses pile up, there are attempts by some scientists to organize these hypotheses in some way in order to bring some structure and simplicity to the field. Theories are formulated in an attempt to understand nature on a deeper level than the individual unrelated hypotheses. Theories propose relationships not only between events, but also between hypotheses, and even smaller theories.

    Often scientific theorists speak of "forming a picture or map" or "fitting the pieces together" and so forth when they try to explain what they attempt to do. Indeed, a very rough analogy could be made with a person trying to put together a jigsaw puzzle. Like the theorist, he is trying to arrange the seemingly unrelated pieces in such a way so as to form a meaningful picture.
    But this analogy cannot be pushed much further. Actually there are

more differences than similarities between the theory builder and the
jigsaw puzzle builder and these differences reveal not only why theory
building is incomparably more difficult, but also why theory building is
actually a completely different enterprise from building a jigsaw puzzle.

One big difference between the two is that the person building the
jigsaw puzzle usually has all the pieces of his picture and therefore
can, in a very mechanical way, slowly build up the picture of the puzzle. The theorist is rarely in such a luckly position. Usually the theorist only has a few pieces of evidence and most of them do not fit directly with any other piece. The pieces can be moved around in a
seemingly infinite number of ways and the theorist must use a good
deal of imagination to propose what might be in the large empty spaces.

Another reason the analogy with the jigsaw puzzle breaks down is
that the jigsaw puzzle can form one and only one picture. Since the
theorist only has a few pieces, it is possible to postulate several different pictures from the available evidence. Turn some of the pieces
around and reorganize them, and several completely different pictures
become possible. Of course as scientists collect more and more
pieces, the total number of possible pictures is slowly reduced, but it
is doubtful if the number of different pictures is ever reduced to only
one.

Also the jigsaw puzzle builder has a tremendously simplified job
because he knows not only that he has all the pieces to his puzzle, but
also he can safely assume that there are no extra or unnecessary
pieces included in his box. A truer analogy with a theory builder's
problem would be if someone took some of the pieces from each of
five different jigsaw puzzles and mixed them all together and then arbitrarily "organized" the pieces by putting all the green pieces together, all the red pieces together and so on. A theorist seldom
makes use of *all* of the available evidence in his field, in fact he is often quite selective, picking only a few pieces from here, a few from
there and so on. A new theory can often be seen as a complete reorganization of facts, which cuts through traditional subject boundaries
and classifications.

Lastly, the scientific puzzle differs from the jigsaw puzzle in that
the jigsaw has definite boundaries. A jigsaw puzzle has four straight
sides and it is clear that the entire picture is contained within those
boundaries. This is not so with scientific theory building. A picture
constructed by a theorist only has fuzzy and jagged edges, there are
never any real boundaries. Even when the theorist has most of the

pieces in his proposed picture, he can never be sure that at some later date more pieces will not be found which will reveal his proposed picture as only one part of a larger picture. Einstein's relativistic theory revealed that Newtonian theory was only a limited proposal--that Newton's theory did *not* hold for all bodies in the universe. Newton's theory provides answers that are approximately correct only when the relative velocities of the bodies are small compared with the velocity of light.

Not only may the theorist's proposed theory be shown to be limited someday, but also it may be downright misleading. Like the blind man trying to describe an elephant by only feeling the tail, the theorist's limited access to the evidence may cause him to propose a completely misleading picture of nature.

No scientific theory can ever be regarded as the complete and final truth. Most, if not all, theories require at least some modification and many are eventually discarded altogether in favour of new theories. A theory is a creative proposal linking together the evidence the theorist has at his disposal. No scientist can ever be sure that a particular theory pictures reality "as it really is". Even if a theory did actually describe nature "as it really is", we would never have any method of knowing this. It is always possible that some new evidence will someday turn up to topple even the most prestigious of scientific theories. A theory must always be regarded as only a provisional explanation, to be replaced if and when necessary. Scientific theories have been discarded repeatedly in favour of a better theory. There is no reason to assume that some if not all of the theories that today's scientists assume to be valid will not someday meet a similar fate.

Although it is not possible to draw a sharp distinction between hypotheses and theories, there are at least two characteristics that at least some theories have, but few if any hypotheses have. Because theories propose relationships between a larger number of events and often fill in the empty gaps with imaginative proposals, theories often make proposals or projections about phenomena that scientists have not yet explored. Reverting back to the crude analogy of a jigsaw puzzle, if a theorist arranges his scattered pieces of evidence in such a way that they seem to form a partial picture of a cat, then the theorist can look at an empty space and propose that there should be a paw here or an ear there and so forth. Experimenters can then draw their attention to this area and test to either confirm or reject this theoretical projection. If these theoretical projections are confirmed by experimentation, this generates a great deal of support for this particular theory within the scientific community.

A good example of one of these theoretical predictions has already been briefly cited in the Preface. Einstein's theory of relativity predicted that the light from stars would be bent as it passed near the sun. Scientists tested this proposal by photographing stars near the sun during a total solar eclipse. The photographs revealed that the light from the stars had indeed been bent as it passed near our sun. Another example of a theoretical projection can be found in Chapter 4 of this book. From my theory of human intelligence, I have proposed that in a particular laboratory situation, humans will be able to learn certain connections which other animals have already been found to be incapable of learning.

A second characteristic which some, but not all theories share is what is known as a "proposed entity". Some theories propose the existence of something that cannot be directly perceived (either directly with our own senses or with the aid of instruments such as a telescope, microscope or geiger counter). Theorists sometimes find that in order to relate the relevant events in a meaningful way, it is necessary to propose the existence of an object or process that can not be perceived even with the aid of instruments. Einstein has used the example of a man trying to understand how a watch works without being able to open the case. After carefully noting the movements of the hands, the ticking sound, the effects of winding or not winding the stem and so forth, the man might well propose the existence of objects he could not see, in this instance supposedly the existence of several gears, springs and so forth in a particular relationship to each other.

It should be pointed out here, that in this situation, there are in all likelihood several different theories which could be formulated which could account almost equally well for all the above movements of hands, sounds and so forth of a sealed watch. Even after scientists had accepted the best of these theories, they would have no way of knowing if the theory actually described the inside of the watch "as it really is". And if more accurate measurement of the movements or sounds became possible in the future, this theory might have to be replaced by another theory that accounted for this new improved evidence.

Actual examples of "proposed entities" from scientific theories include electricity, molecules and genes. This subject of proposed entities will be discussed in more detail in Chapter 8.

Although similar in nature to hypotheses, theories usually attempt to relate more events and sometimes include theoretical projections and proposed entities. In the early stages of a new science, scientists proceed largely by "blind" experimentation--by proposing individual, almost random hypotheses and then testing them. After a new science has accumulated a certain number of supported hypotheses, inevitably a few scientists attempt to formulate a theory which brings these seemingly unrelated events and hypotheses into a "picture" or "map", that provides a larger and more comprehensive understanding of the problem.

A scientist proceeding by "blind" experimentation can be compared to one of Thorndike's cats proceeding by "blind" trial and error. Blind experimentation is the separated equivalent to contiguous blind trial and error learning. And hence theory building can be seen as the separated equivalent to contiguous insight learning. Rats are capable of learning contiguous relationships through blind trial and error, but they are not able to learn separated relationships through the use of hypotheses and blind experimentation. Likewise, chimps are sometimes capable of insight learning, that is seeing the essential relationships between objects when these objects are all visible at the same time, but only man is capable of theorizing--of proposing a relationship between many separated events.

In Chapter 3 it was concluded that it is not possible to draw a clear distinction point between trial and error learning and gestalt insight learning. Likewise, it is not possible to make a sharp line between blind experimentation by random hypotheses and theory building, the one merges into the other.

Often in a new science there are several competing theories which attempt to relate much of the same phenomena in different ways. These competing theories are often called "schools". Older sciences such as astronomy and mathematics had this early period in prehistory, whereas some of the newer sciences such as psychology and anthropology are still in this stage. Usually one of these competing theories or a new enlarged theory is gradually accepted by most scientists in the field to be the best proposal at the present time.

Thomas Kuhn (1970) has described what happens after one theory has been accepted by almost all, if not all the scientists in a particular field or sub-field. This theory is said to be a "paradigm" a basic dominant theory that is assumed by the scientific community in the field to be valid, at least until a better alternative is proposed. The science is then said to be a mature science. Although strenuously opposed by many doctors when first introduced over a hundred years ago, germ theory can now be said to be a paradigm in the medical scientific community.

After a young science has received its first universally recognized theory or paradigm this paradigm is usually felt to account for and explain quite successfully the relevant evidence and observations. Inevitably scientists assume that this theory is indeed valid. Often this paradigm* is accepted for many years, even centuries, by scientists.

A "scientific revolution" occurs when a new theory overthrows an older paradigm and itself becomes the paradigm. Examples of such scientific revolutions occurred when the Copernicus theory of a sun-centred system overthrew Ptolemy's theory of an earth-centred universe, or when Einstein's theory of relativity replaced Newton's theory of universal gravity.

Once a science has arrived at the mature stage, the work of the scientists becomes what Kuhn has called "normal" science and much more closely resembles the activities of the jigsaw puzzle builder rather than the tremendously more complicated creative reorganization of evidence proposed by the theory builder. A scientist working within a paradigm in effect assumes that this theory describes what the world is really like. In terms of our jigsaw analogy, the scientist assumes that this problem is indeed a partial picture of a cat and concerns himself solely with working out the details.

In most cases there is a great deal of "mop-up" work to do on a particular paradigm, such as careful description and measurement of the relevant events, trying to expand the boundaries of the picture, looking for the missing pieces and so on. Many scientists spend their entire scientific careers on this "mopping up". Scientists working within a particular paradigm inevitably assume the theory to be valid and are much more likely to propose only new hypotheses that fit nicely within that theory. As a consequence, these scientists often ignore or suppress the few pieces of evidence that refuse to fit into the picture. The paradigm has been so successful in the past, that the scientist assumes that these few "difficult" pieces will eventually fit in somewhere or that they are errors in measurement and so forth.

Normal science is committed to a particular paradigm. Scientists working within a paradigm usually do not aim to invent new theories and are often intolerant of new theories proposed by others. The edu-

---

*In response to critics who claimed that Kuhn was using the word "paradigm" in more than one sense, Kuhn has recently proposed that the term paradigm be sub-divided into "disciplinary matrix" and "exemplars". Readers wishing to pursue this distinction see Kuhn, 1977, p. 293-319.

cational process which all aspiring scientists must traverse is almost
always narrow and rigid. Scientific training is designed to create a
scientist who will work smoothly within the given paradigm, not to
produce a creative and disruptive theorist.

Nevertheless, scientific revolutions have occurred many times when
at least one scientist, after a great deal of thought about the few pieces
that have repeatedly failed to fit in the proposed theory, decides that
this proposed picture cannot account for these difficult pieces and after
juggling the pieces around in many different combinations, finally
decides that a new proposal can better accommodate all the evidence.
In our jigsaw example, he may propose that this is not really a picture
of a cat at all, but rather a picture of a lady dressed in a mink coat!

One of the best examples of how a new theory can completely reorganize the evidence is Copernicus's theory of a sun-centred or helio-
centric system. Since the time of Aristotle, early science had accepted
the earth-centred universe theory. This is the most obvious and "com-
mon-sense" approach, certainly the earth appears to be stationary,
the sun appears to orbit around the earth once a day. The moon,
planets and stars also appear to move around the earth. It seems obvious that the earth is fixed at the centre of the universe, with all the
heavenly bodies revolving around it.

This earth-centred concept was modified and perfected by Ptolemy
of Alexandria in the second century A.D. This modified system was
able to account for some of the more obvious weaknesses of an earth-
centred system. For example, although the sun, moon and stars
moved as predicted from this theory, the planets did not. Planets'
paths sometimes, apparently, temporarily reverse their direction of
motion. The word planet itself comes from the Greek word for
wanderer. The modified theory took into account most of these "difficult" facts by proposing a complicated series of three simultaneous
planetary circular motions.

Through the centuries whenever a new discrepancy appeared, scientists would simply eliminate it by making a minor adjustment in
Ptolemy's system of compounded circles. However, by Copernicus's
time in the early part of the sixteenth century, the system had become
extremely complicated, requiring more than 70 simultaneous motions
for the seven celestial bodies.

Copernicus finally decided that all the evidence could be better
accounted for by proposing a completely different theory. He proposed
that the sun and not the earth was at the centre, and that the planets
and earth revolved arount it.

Ptolemy's theory proposed a picture of the universe with the earth
at the centre. The theory accounted for most of the evidence, but there
were always a few difficult events that did not seem to fit in properly.
Copernicus completely reorganized the evidence, proposing a new picture with the sun at the centre. His theory was much simpler and was
eventually found to accommodate the evidence better.

Usually these new theories are proposed by either young scientists
usually thirty or under or else by older people who have only recently
entered the field. Presumably this is the case because these people
have not been steeped too long in the older theory and are still willing
to try out different alternatives.

Needless to say, after a new theory which completely reorganizes
the available evidence has been proposed, there is a great deal of
conflict within the scientific community. The defenders of each theory
claim their theory can best account for the evidence. There is usually
a great deal of work on the few areas where the two theories directly
conflict and it is possible to test in these "crucial experiments" to
support either one theory or the other.

Often the scientific community takes many years to switch from the
old paradigm to the new. Often there is only a slow trickle of converts
from the old paradigm to the new. It seems particularly hard for a
scientist who has spent his entire life working in one paradigm to
switch over to the new paradigm. Often the young scientists accept
the new paradigm, and a change over in the community is not complete
until all the older scientists have died.

Kuhn challenges the standard view that science develops by the gradual accumulation of discoveries and inventions. The existence of scientific revolutions shows that perhaps the most important and exciting
aspects of science are the non-cumulative developmental episodes.
New theories seldom if ever add any new evidence, instead they reorganize the evidence that is already at hand.

FORESIGHT

Humans are the only animals capable of separated learning. Other
animals are limited to contiguous learning, but we are capable of both
contiguous and separated learning. Given this unique ability, what particular advantage does this give us over the other animals? How does
this ability improve our chances of survival?

Before examining this it will first be necessary to briefly discuss
the problem of "cause and effect". I have avoided using this term until now and I have even tried to eliminate the word "cause" as much as

was possible. This is a very old problem in the philosophy of science which was first brought to light over 200 years ago by David Hume (1739).

In essence, what Hume said was that we really do not know what "causes" things to happen. As Hume put it, "The powers by which bodies operate, are entirely unknown." After observing that one event always follows another event, someone may propose that the first "causes" the other. But this is not a justified conclusion. It may be that the second event only happened to follow the first event when you were observing. Even after this sequence of events has occurred hundreds or thousands of times, it can never be said the one event *causes* the other.

Consider the rat in his proverbial Skinner box. After observing hundreds of times that the food pellet always follows the pressing of the lever, the rat might well conclude that the one invariably *caused* the other. But the next day when he presses the lever, a food pellet may not appear. The pressing of the lever does not invariably *cause* a food pellet to drop.

Similarly, we have observed apples and other objects fall to the ground thousands of times and it is very tempting for us to conclude that the earth *causes* objects to fall to it. But like the rat, we may someday be shown otherwise. It is possible that tomorrow an apple will not fall to the ground. It might just hang there in mid-air or fly off in the opposite direction or who knows what.

Like the rat in his Skinner box, we are totally ignorant of the mechanism that produces connections between events. To repeat Hume's phrase, "The powers by which bodies operate, are entirely unknown." We observe and note objects' movements, but *why* objects react in those ways we have no knowledge whatsoever. We are accustomed to thinking that "objects fall to the ground because of gravity". But this is really a circular statement since gravity is defined as the attraction between objects. This results in the statement that objects fall towards each other because they attract each other. The stark truth' is we do not know why objects attract each other.

Earlier it was stated that modern germ theory holds that micro-organisms "cause" infections. By this doctors mean that a particular germ is always present in a patient with a particular disease. A certain set of symptoms and a particular micro-organism always go together. But doctors still do not know *why* a particular germ causes a particular disease. There is always some point in a scientist's explanations when he comes to a stop--when he must say, "I do not know why these events are related to each other."

As I said before, the whole universe is our Skinner box. We observe objects in our universe and are able to learn that connections appear to exist between particular events, but we do not know why these connections exist. We are most fortunate that at least so far, most of the connections do not change from day to day. There is not a single verified recorded instance of an object not being attracted to the earth. The poor rat may find that the "laws" of his box change from day to day. One day a pellet falls each time he presses the lever, the next day he finds that he must press it five times before a pellet falls, and the next day no pellets may fall no matter how many times he presses the lever and so on. It is possible that some day the "laws" of our universe will change as well.

The term "law" has fallen into disrepute among scientists. For one thing it implies not only that things have acted this way in the past, but also that they are required to act the same way in the future. The term also seemed to imply that scientists have discovered the final truth about nature. At one time a "theory" was supposed to be only a proposal about nature and a "law" was a theory that had been proved to be valid. This of course is no longer tenable, Newton's "universal law" of gravity is now accepted as only a limited theory.

Since we do not know *why* connections exist between events, we also have no way of knowing if the same connections will continue to exist in the future. We *invariably assume* that the connections that existed yesterday will also exist tomorrow, but we actually have no valid reason for doing so. Both man and also the other animals make this assumption. After the rat has pressed the lever, he runs over to the food tray, obviously assuming that a pellet will shortly appear. But he is sometimes wrong.

Hume, of course, also made this assumption and realized that everyone else did as well. Although this is an ungrounded assumption, people have always made it and will continue to do so in the future. Indeed in order to survive, this assumption must be made. Hume was only trying to show that this *is* an assumption on our part and not, as had previously been thought, a quality of nature itself.

The reason for bringing this subject up here is that it is important to note that Hume's two major points: (a) we do not know why connections exist between events, and (b) we therefore have no valid reason for assuming that these same connections will continue to exist in the future, hold equally for both contiguous connections and also separated connections. From previous experience, we assume that when we stick our hand in a roaring fire, we will almost immediately feel pain; we also assume that if we let dirt get in a wound, the wound will probably become infected in a few days. The first example is a contiguous connection, the second is a separated connection, but both involve the same underlying assumption that the connections we have learned in the past will continue to exist in the future. Hume's assertions hold for both contiguous and separated connections.

This brings us back around to the actual theme of this section, what particular advantage does our unique separated learning ability give us? Remember that it has been previously discussed in Chapter 3 that all animals are capable of some degree of "foresight", that is seeing what will probably happen in the future. After an animal has learned a particular connection it is obvious from his actions that he expects the same connection to exist in the future.

In trial and error learning, after an animal has learned a particular connection, the animal certainly acts as if he expected this same connection to exist in the future. Thorpe (1962) has called this the "principle of expectancy". Although all animals have some limited amount of expectancy or foresight, psychologists agree that humans have it to a far greater degree.

However, with the discovery of separated learning, it is now possible to show exactly how the foresight of humans differs from that of other animals. Foresight in all other animals is limited to not more than 60 seconds. Since all other animals are limited to learning contiguous connections--that is those in which not more than 60 seconds separates the events--their foresight is also limited to not more than 60 seconds.

Foresight is defined here as anticipating from previous experience what events are likely to occur in the future. Once a rat has learned that a food pellet always falls a few seconds after he presses the lever, he has some foresight about what will probably happen the next time he presses the same lever. But since he is limited to learning only contiguous connections, his foresight is also similarily limited.

Humans are not so limited in foresight however. Since we are capable of learning separated connections, we are also capable of foresight that extends considerably beyond one minute. To make the terminology as simple as possible, we will say that all other animals are limited to *contiguous foresight*, that is, foresight that results from contiguous learning and that only humans are capable of *separated foresight*--foresight resulting from separated learning.

All foresight in humans and in other animals is based on the ungrounded assumption that the connections which were learned yesterday will continue to exist tomorrow. But as Hume pointed out we have always made this assumption, we always will, and indeed we must do so in order to survive.

I am equating human or separated foresight with "scientific prediction". After a scientist has discovered a particualr separated connection, he can often predict what is likely to happen in the future. This separated or scientific foresight is of course seldom if ever 100% accurate. There are at least two possible reasons for this; the hypothesis or theory may be faulty, inaccurate or incomplete or the connection might change in the future. When scientists find that their predictions are not completely accurate, they usually start revising their hypotheses or theories.

The extent of scientific foresight depends on the interval between the relevant events. A man that had just been bitten by a yellow fever-infected mosquito would have some foresight about what would probably happen to him in three to six days. A man who had contracted syphilis (and did not seek medical care) would have foresight about what would be likely to happen to him in twenty to thirty years. The longer the interval between the connected events, the longer the foresight into the future.

In a mature science, predictions are often quite accurate. At the present time astronomy is easily the science with the most accurate and long range predictions or foresight. Astronomers are able to predict what will happen hundreds of years in the future. For example, astronomers now predict that there will be a total solar eclipse on 17 November 2161 over southern Africa and Madagascar (from Oppolzer, 1962). This is foresight approaching two hundred years into the future. Scientists could forecast eclipses much further into the future if they wanted to.

Not all hypotheses or theories allow accurate foresight or prediction, however. It depends on the nature of the proposed relationship and how specific the proposal is. One example is the hypothesis that "smoking cigarettes may result in lung cancer several years later". Even assuming for a moment that this hypothesis is valid, the hypothesis only proposes that one event may be related to the other in some cases. Nor does the hypothesis propose a precise interval between the events. In its present form, this hypothesis does not allow a doctor to predict which cigarette smoker will be likely to have lung cancer in X number of years.

Theories also vary in the amount of foresight they allow. Although

Darwin's theory of evolution relates thousands of events, it is not specific enough to allow a biologist to predict exactly when a given species will become extinct or when a new species will come into being. The theory of evolution only proposes in general terms that a given species will continue to survive only as long as it successfully adapts to its environment.

The gist of all this is that only humans are able to guess what is likely to happen more than a minute into the future. Through secondary learning other animals sometimes appear to have foresight of up to several minutes into the future, but this cannot be regarded as genuine foresight. As discussed in Chapter 3, secondary learning is actually only contiguous learning and hence only appears to be separated learning. Therefore secondary learning results in only apparent separated foresight. Even with secondary learning, the apparent separated foresight of other animals extends not more than a few minutes into the future.

No other animal is capable of realizing what will probably happen in an hour from now, tomorrow, next week or next year. It follows that other animals must live only from moment to moment, they only react to their immediate sensations. One is tempted to use the phrase "out of sight, out of mind", but this is misleading because it also implies that other animals have a very short memory. We know from experiments that some animals have a quite long memory. But the ability to remember what happened in the past is not the same as the ability to realize what will perhaps happen in the future. Foresight depends on the ability to learn that certain connections exist between events. Memory requires no such learning. Other animals may have a long memory, but they have a very short foresight.

Even chimpanzees, the animals generally considered to be the most intelligent of the non-human animals, do not show evidence of separated foresight. Wolfgang Kohler, the gestalt psychologist who pioneered the study of chimp learning and intelligence, concluded in his classic book *The Mentality of Apes*(1927), that:

> the experiments in which we tested these animals brought them into situations in which all essential conditions were actually visible, and the solution could be achieved immediately . . . we /di̲d̲/ not test at all, or rather only once in passing, how far the chimpanzee is influenced by factors not present, whether things 'merely thought about' occupy him noticeably at all. . . we have not been able to tell how far back and forward stretches the time 'in which the chimpanzee lives'. . . A great many

years spent with chimpanzees lead me to venture the opinion
that besides in the lack of speech, it is in the extremely narrow
limits in *this* direction that the chief difference is to be found
between anthropoids and even the most primitive human beings
(p. 266-267).

By now it should be clear what particular advantage man's unique
separated learning ability gives him. Our separated foresight is of
tremendous survival value to us. After we have succeeded in discovering a separated relationship between two events, we have some foresight about what will probably happen in the future after the first event
has occurred. If the second event is an event which is favourable to
our survival, then we can attempt to increase its occurrence or if it is
not helpful to our survival, we can either try to decrease its occurrence or at least reduce its harmful effects.

Since childbed fever is of course not helpful to our survival, when
Semmelweis found that "cadaveric matter" in open wounds was related
to childbed fever two to three days later, he tried to prevent cadaveric
matter from entering open wounds. Without separated foresight, Semmelweis could not have taken such action. Even if it is not possible to
stop the first event or if the event is not "causal", our separated foresight can still be useful. For example, if it turns out that the changes
in "seismic velocity" of rocks are indeed related to earthquakes, our
separated foresight will allow people to evacuate a city just before an
earthquake and save many lives.

Since other animals are limited to contiguous foresight, they are
limited in changing their individual behaviour towards only those events
that happen to have contiguous connections. A human is able to adjust
and adapt his behaviour to suit not only events that are likely to occur
in the next few seconds, but also events that are likely to occur days,
weeks or even years from now. Exactly how humans adapt their
behaviour in the light of separated foresight will be discussed and
explained in Part Three.

As our storehouse of knowledge of separated relationships becomes
greater and more accurate, our separated foresight also becomes
greater and clearer. Thanks to the growth and development of science,
our foresight about future events is much more extensive and accurate
than that of our own grandfathers, much less the extremely limited
foresight of primitive man. Human foresight is not completely accurate, and never will be, but even a foggy and blurred foresight of
events that may happen far in the future has a tremendous survival
value.

Chapter 6

Apparent Contradictions

Before continuing the discussion concerning our unique learning ability, it might be best to stand back in this chapter and consider some of the objections which will surely be raised to the proposals I have just outlined. There are at least two main objections which the knowledgeable reader in this field may have found to my proposal that humans are the only animals capable of separated learning and separated foresight. The first objection stems from a recent series of experiments which has shown that rats and other animals are able to learn to avoid poisonous foods. After a rat has eaten a small amount of a poisonous food, he sometimes does not become sick until several hours later. When the rat subsequently refuses to eat this food, this shows that he has learned an association between this particular food and his sickness several hours later. In other words, this learning by the rat was not contiguous.

The second objection that could be raised to my proposal that humans are unique in separated foresight stems from the behaviour of some animals which prepares them for events that will occur several weeks or months in the future. The migration of birds is a good example. Birds fly south in the autumn obviously in order to avoid the cold weather which will arrive in a few weeks time. It might be concluded that these birds have some "foresight" about the forthcoming winter. These two objections will be considered in this chapter and it will be explained why they are only apparent contradictions to our unique separated learning and foresight ability.

TASTE-AVERSION LEARNING

In Chapter 3 it was concluded that with only two very narrow exceptions, all learning among non-human animals is contiguous. The second limited exception will now be considered. Actually this exception has only recently been discovered--all the important experiments in this area have been conducted within the past few years.

Rats obtain virtually all their food and shelter from humans. They live mostly in and under our buildings and eat our wasted or unprotected food. Humans have tried for centuries to rid themselves of this pest, but have obtained very little success. Rats are very adept at avoiding our poison and traps.

Our constant frustration in attempting to destroy the rat has sometimes led men to attribute to them a high degree of intelligence. One scientist called the conflict between men and rats "a veritable battle of wits". However, careful observation and experimentation on rats has shown that their avoidance of strange objects is indiscriminate, it applies to not only harmful objects, but also to virtually any strange object. Their avoidance is automatic rather than specific.

This very useful reaction of rats, which helps to protect them from our extermination attempts, has been called the "new object reaction". This behaviour has been described by Barnett (1963). Wild rats (but not tame laboratory rat strains) will completely avoid for days or sometimes weeks any new or unfamiliar object which is placed in one of their pathways. They will even avoid a heap of wholesome food for a time. Apparently, this avoidance reaction is limited to strange objects in an otherwise familiar situation. In a totally strange environment, rats will thoroughly investigate the area.

After the "new object reaction" fades, rats continue to be very cautious of any new food. When they do begin to eat a new food, they by no means eat their fill. On the first day they only eat a very small sample, the next day a little more and so on until finally they eat it freely. Since they eat only a small amount on the first day, if the rat food is poisonous, they have a fairly good chance of recovering. A rat that has sampled a small amount of poisoned food will refuse this same food on subsequent occasions. This feeding behaviour is extraordinarily effective in protecting rats from our poisoning.

The ability of rats to learn to avoid poisonous foods may be very useful for the survival of the rat, but it has recently caused a lot of headaches for psychologists studying learning. As discussed before, psychologists have long held that all learning is contiguous. But even when rats were fed poisons that took several hours to act, they were still able to learn to avoid the poisonous food. The rat's learning was non-contiguous. Also, this at first appears to be an exception to my proposal that humans are the only animals capable of separated or scientific learning.

The first laboratory experiment in this area was conducted by Garcia, Kimeldorf & Koelling in 1955. At first there was a great deal of scepticism by learning psychologists concerning the results of these early experiments. Nevertheless, these findings have been replicated

many times in the past few years. A recent review of the work done
in this area can be found in Logue (1979).

Experimenters have studied many different aspects of this type of
learning, so that now we are beginning to have a little better understanding of its limits and nature. Although a few experimenters have
actually fed their rats poisonous or toxic drugs (such as lithium chloride) in their food or water, in order to facilitate greater ease and
control in experimentation, most of the studies have fed the rats a
distinctively-flavoured water (such as salt- or sugar-flavoured water)
and then later artificially produced gastric upset in the rats by either
exposure to X-rays (causing "radiation sickness") or by hypodermic
injections of a drug such as apomorphine. By artificially producing
an upset stomach in the rat, psychologists have been able to vary the
interval between the ingestion of the food and the beginning of the
gastric upset. Etscorn & Stephens (1973) found that rats were able to
learn aversions with up to a 24-hour interval between consumption and
sickness.

By giving the rats two differently flavoured solutions, either at the
same time or else one following the other by a few minutes, experimenters have been able to uncover more information about this type of
learning. For example, they have found that rats learn aversions to
new or novel solutions much more readily than they do to familiar solutions. Also, solutions which have a bitter taste are more likely to
be associated with the aversive consequences. There is also a tendency for the rat to learn a stronger aversion to the last solution
consumed.

For our purposes, however, the most important finding in this area
is that although this is non-contiguous learning, it is a very limited
and rigid type of learning. Rats are only able to learn non-contiguous
aversions to food or water. This type of learning can occur only between a food-related event (such as taste, to a lesser extent smell or
to an even lesser extent the sight of the food or container) and gastrointestinal upset. It has been shown repeatedly that rats are unable to
learn non-contiguous aversions to any other type of event such as
sound, location, light level or solution texture. This non-contiguous
learning is limited to aversions resulting from an upset stomach, it is
not possible with any other "punishment" (such as electric shock).
And only if a rat's upset stomach was caused by something he ate or
drank, will he be able to avoid it the next time. If it was caused by
any other reason, he will not be able to develop an aversion to it.

Although in rats and most other animals this type of learning is

mainly limited to the taste or to a lesser extent the smell of food, at
least one species is easily able to use visual cues to learn to avoid
poisonous foods. Wilcoxon, Dragoin & Kral (1971) found that the bobwhite quail, which uses its excellent visual system to direct its pecking
responses, was able to develop non-contiguous aversions to blue col-
oured water. The bobwhite quail uses mostly visual cues to identify
its food, whereas most animals rely primarily on taste and smell. In
all animals this type of learning is limited to food-related events
(which may vary from species to species) and gastric upset.

Particularly at first, there was a great deal of reluctance by learning psychologists to accept that this was actually non-contiguous learning. Several *ad hoc* theories were proposed to explain these results
as only apparent non-contiguous learning. For example the "aftertaste
theory" proposed that in some way an aftertaste or regurgitation or a
high concentration level in the blood could be responsible for this
learning. In other words, that an aftertaste or such was still present
in the rat's mouth at the onset of the gastric upset, thus creating
temporal contiguity.

Experimental results have not supported this idea however. Rats
are able to learn an aversion to a specific concentration of a solution
as opposed to other concentrations of the same solution. They can also
learn an aversion to a particular solution even if other solutions are
drunk between the first solution and the stomach upset. It is difficult
to imagine any aftertaste after one or more other solutions have been
consumed. One study even found that rats learn an aversion to a particular temperature of water. It is highly unlikely that rats could have
an aftertaste of temperature. Also, rats seldom if ever regurgitate
because their regurgitation is blocked by a cardiac sphincter.

Findings from many experiments seem to show that this is indeed
non-contiguous learning. The non-contiguous aversions can not be
conveniently explained away. Although non-contiguous, this type of
learning is very narrow, only food-related events can be associated
with stomach upset. Psychologists have only scratched the surface in
the understanding of this type of learning--there is much more to discover concerning this type of non-contiguous learning. But it is already clear that it will have to be classified as yet another type of
learning.

Why is taste-aversion learning so narrow and limited? We do not
really know, but part of the answer may be that with this form of learning, animals do not test or try out various solutions as they do with

other forms of learning. Remember that with instrumental or "trial
and error" learning, animals often appear to be making systematic
attempts at solution to their learning problem. First they try one solution, then if this doesn't work, they try another and so on. However,
with taste-aversion learning they do not do this. For example, when
presented with two differently flavoured fluids followed by illness, the
rats never try or test first one, then the other fluid in an attempt to
find out which one made them sick.

Kalat & Rozin (1972) conducted an experiment in which they tried to
help the rats out as much as possible in learning which of two fluids
had made them sick. The rats were first given salt-flavoured water,
then a few minutes later sugar water, and then after a few more minutes were intubated with a toxic drug. This resulted in a mild aversion
to *both* solutions. Then some of the rats were extinguished to the sugar
solution, that is they were given sugar water several times without any
accompanying poison. The rest of the rats were not so extinguished.
Finally all the rats were presented with a choice between salt water
and plain tap water.

In other words the rats first experienced an ambiguous situation in
which either salt or sugar-flavoured water could have made them sick.
Then some of the rats were allowed to learn that the sugar water was
"safe" (it did not make them sick). In short, Kalat & Rozin had forced
some of the rats to "try" or test the sugar water. They then reasoned
that if the rat's learning was an advanced, or as they call it "cognitive"
type of learning (i.e. separated or scientific learning), the rats would
be able to think "I now know it wasn't the sugar water, so it must have
been the salt water." If this was the case, the rats that had been extinguished to the sugar solution would drink far less salt solution than
the rest of the rats that had not been extinguished. This however was
not the case, as both groups drank about the same amount of salt
water.

So even when the rats were literally forced to "try" or test one
possibility, they were unable to benefit from this experience. This
may partially explain why taste-aversion learning is so limited and
narrow in its range. This type of non-contiguous learning only occurs
between a food-related event and gastrointestinal upset. But even
within this narrow range, if presented with more than one food-related
event, they are unable to test or try out various solutions. Taste-
aversion learning is a narrow, rigid, even "automatic" form of learning where no flexibility exists because there is no "trial and error"
(trying out various possibilities).

This explains why taste-aversion learning does not contradict my proposal that only humans are capable of separated learning. Separated learning was defined (p. 99) as learning by trial and error or experimentation that a connection exists between two or more events which are separated by time. In this context, the important phrase in the last sentence is "learning by trial and error or experimentation". Scientists of course are continually testing and trying out various solutions to their problems. As a result science is an extremely broad, general and flexible form of learning. Scientific "experimentation" is of course, only a very sophisticated form of trial and error learning. One possible solution is tried (tested) and then another and so on.

Animals use trial and error in learning contiguous relationships, but they do not do so in taste-aversion learning. Humans, on the other hand, use trial and error or experimentation for both contiguous and non-contiguous learning. Both taste-aversion learning and scientific learning are non-contiguous, only science proceeds by trial and error.

There is more evidence that taste-aversion learning is a completely different type of learning from science. This evidence suggest that perhaps the taste does not become a signal for the poison in the same way that the lever becomes a signal for the electric shock. In taste-aversion learning, the perception of the taste itself may change and become aversive. In other words, instead of the rat looking at the food and thinking, "That looks good, but if I eat it I will get sick in a few hours", there is some evidence to suggest that the rats may look at the food and think "Ugh, I don't fancy that food any more".

Although most of the experiments in this area have been conducted with rats, this type of learning is by no means limited to this animal. It appears that most, if not all vertebrate animals including humans are capable of this type of learning. It may be however, that rats are more adept at this type of learning owing to the selective pressure of centuries of attempted poisonings by humans.

Many people have experienced this type of learning. After becoming sick at a party, they find that they have suddenly "gone off" one of the foods or drink that they had consumed at the party. Although we of course do not really know what a rat thinks or feels when he refuses to eat a food, humans at least are able to tell us why they do not want a particular food any more. These human reports support the view that in this type of learning, the taste itself becomes aversive or unacceptable. What was formerly a well-liked food suddenly becomes unpalatable.

Martin Seligman (in Seligman & Hager, 1972, p. 8) gives an account

of his experience with taste-aversion learning. Seligman explains that
a particular sauce (Sauce Bearnaise) used to be his favourite. One
night he and his wife went out to dinner and ate filet mignon with this
sauce, and about six hours later he became violently ill and spent most
of the night vomiting. The next time he had the sauce, he couldn't
bear the taste of it.

The taste of this sauce was the only thing that had become aversive.
Neither the restaurant where they ate, nor the white plates off which
they had eaten the sauce, nor the opera they had listened to after the
dinner, nor even his wife who had been with him all this time, became
aversive. Only the taste itself of this particular sauce was regarded
differently.

If all animals are capable of this limited taste-aversion learning,
and if only humans are capable of separated or scientific learning, it
follows that only humans and then only when concerned with upset
stomach, are capable of both of these types of non-contiguous learning.
This is in fact the case. Following an upset stomach, humans are
capable of both types of learning. For example, when many people at
a dinner or party become sick, we are often able to discover which
food was the culprit by the use of scientific learning. Each person is
asked exactly what he ate, and by finding the one food which all the
sick people ate, (but none of the other people ate), it is possible to
propose a hypothesis about which food caused the illness. With scientific learning, the taste of the food itself does not change--the food may
still look appealing--but the person now knows that if he eats it he will
probably become sick in a few hours time. The net result of both types
of learning is the same--the person refused to eat a particular food--
but the internal mechanism of the learning is different.

Since taste-aversion learning is limited to developing aversions to
only food-related events, it follows that this type of learning can often
be faulty. If a person becomes sick in the stomach for any other reason besides something he ate or drank, taste-aversion learning may
cause him to become aversive to a totally wholesome food. This is in
fact what happened to Seligman. Using scientific learning, Seligman
concluded that the sauce was not actually the cause of his illness.
Since his wife had also eaten the sauce and did not become sick, and
since he later found out that one of his follow workers at his laboratory
had also had a similar illness at the same time, Seligman concluded
that his sickness was actually caused by the stomach flu and not the
sauce after all. By conducting this "thought experiment", separated
learning led him to hypothesize that it was stomach flu, taste-aversion

learning on the other hand produced an aversion to the sauce. The two types of learning produced two different answers. Presumably taste-aversion learning was at fault, although it is of course always possible that both answers were incorrect. Apparently scientific learning does not cancel out the taste aversions, even after Seligman had decided his illness was actually caused by the stomach flu, the mere thought of the sauce continued to nauseate him for about five years.

Taste-aversion learning can often be faulty, particularly when the stomach illness was not caused by food or drink. This is the case not only in humans, but also in other animals. In the experiments conducted with rats reviewed in the first part of this chapter, most of the experimenters made the rats artificially sick by the use of X-rays or drugs. The rats developed non-contiguous aversions to the food or water, but not to the X-ray chamber or the hypodermic injections. Since the flavoured solutions did not actually cause the rats' illness, it is questionable if the aversions the rats developed in most of these studies could properly be called "learning". At best this could only be called "faulty learning".

Taste-aversion learning can be distinguished from separated or scientific learning. Taste-aversion learning is a narrow, rigid even "automatic" form of learning whereas scientific learning is a very general and flexible form of learning using trial and error or experimentation. Apparently most if not all animals are capable of taste-aversion learning, whereas only humans are capable of separated or scientific learning.

If taste-aversion learning results in a change in the taste itself of a particular food, it follows that taste-aversion learning does not produce true separated foresight. The animal does not realize that eating a particular food will cause him discomfort in a few hours time, he only realizes that this food tastes bad to him *now*.

The problem of learned aversions to poisonous foods appears to be closely related in some ways to the ability of animals to select a balanced diet. Richter (1943) found that given a selection between a large variety of different foods, rats were able to self-select an extremely well-balanced diet. Davis (1935) conducted a similar experiment with human babies only eight to ten months old. The babies were also able to self-select a balanced diet. Although at any one meal the babies might not select a balanced diet, over a period of time the diet was what any scientist would accept as well-balanced.

Scientists do not yet have an explanation for this remarkable phenomenon. When an animal becomes deficient in a particular nutrient, he somehow "knows" he is deficient and is able to choose a food which will remedy this deficiency (assuming of course that the food is available). At least in the case of salt (sodium chloride) deficiency, it appears there is a genetically preprogrammed preference for this substance. Rats which were made salt deficient for the first time showed a strong preference for salt immediately upon tasting it.

It is questionable, however, whether animals could have an innate preference for each of the many different vitamins, minerals and nutrients which they need. There is some evidence that animals are able to learn preferences for the taste of foods needed to recover from specific deficiencies. Careful experimentation has suggested that often animals actually learn an aversion to a deficient diet rather than learning a preference for the specific food containing the necessary nutrient. Deficient rats show an immediate preference for new foods. When given a choice between two different novel foods, only one of which was enriched, deficient rats did not quickly develop a preference for the enriched diet. See Rozin & Kalat (1971) for a review of work in this area.

Learning an aversion to a deficient diet and learning an aversion to a poison are obviously similar. A deficient diet can be seen as a slowly acting poison. Exactly how this type of learning operates is not yet known. There is some evidence for assuming that the animal's perception of the taste of particular foods changes over time as chemical deficiencies develop or after a gastrointestinal upset. Richter (1943) found that rats which had their taste nerves cut were no longer able to select a balanced diet. Also humans that have experienced a deprivation of a particular nutrient often develop a strong craving for a food that contains the necessary nutrient. This evidence supports the hypothesis that the perception of the taste itself changes in this type of learning.

INSTINCTIVE BEHAVIOURS

When Europeans plan their winter holidays, they usually plan to travel south to warmer climates. Spain and Greece are favourite winter resort countries in Europe. Through the use of separated learning about the length and onset of earth's weather seasons, humans have some foresight about what the weather will be like in northern Europe next December. Given this separated foresight of future cold weather, some people plan to head south in the winter to escape some of this uncomfortable weather.

When we observe many species of birds migrating south each au-

tumn, it might be tempting for humans to jump to the conclusion that
these birds also "know" that winter is coming and are leaving now before the cold weather arrives. If many humans travel south in the
winter as a result of separated foresight, it is tempting to conclude
that birds flying south are also capable of separated foresight. However, similar behaviour in different species does not necessarily mean
that the internal mechanism is the same in both species. It is possible for several species to display similar behaviour for different
reasons.

There are many more examples in nature where animals appear to
be preparing for the distant future. Some of the behaviours of many
animals is adaptive and has survival value only in relation to events
which will occur to them days, weeks, sometimes even months in the
future. Not only in the migration of birds and fish, but also the hibernation of some rodents during the winter, nest-building by wasps,
birds and rats, and the hoarding of food by squirrels and other animals
are just a few other examples of animal behaviour that show preparation for the distant future.

A great deal of animal behaviour is concerned with attempting to
ensure the animal's survival in both the immediate and distant future.
Almost all animals are capable of contiguous associative learning,
which is limited to learning connections between events that are separated by only a few seconds. This contiguous learning results in contiguous foresight which allows the animal to "see" only a few seconds
into the future. Most vertebrate animals are also capable of taste-
aversion learning which allows the animal to learn aversions to poisonous foods which may take several hours to act. However, I have
proposed that only humans are capable of separated learning and hence
separated foresight. We are the only animals that are able to know as
a result of experience what will probably happen tomorrow or next
week or next year. If this proposal is valid, how is one to explain all
the activities of animals--such as the migrations of birds--that are
obviously preparing the animal for an event which will occur weeks
in the future.

Even though these behaviours are preparing these animals for
events that will occur in the distant future, these behaviours are not
necessarily *learned*. These activities can be accounted for by what at
various times has been called "instinctive" or "innate" or "genetically
determined" or "fixed action patterns" or "stereotyped" or "species-
characteristic" behaviours. As this multiplicity of terms may suggest,
this is currently an area of conflict and dispute by ethologists and psy-

chologists. This type of behaviour will now be discussed.

It is possible that the idea of "instinct" was one of the earliest concepts formulated by humans to explain animal behaviour. The modern study of instinct began with Charles Darwin (1859). Darwin thought of instinct mainly as an inborn or innate urge for behaviour which is adaptive for the animal's survival, but without the animal knowing the purpose of the activity. The history of the concept of instinct in recent times shows a remarkable variation in acceptance by the scientific community. With the rise in the study of learning during the first part of this century, the concept of instinct was increasingly ignored and finally by the 1920s almost completely abandoned and considered disreputable by most psychologists, zoologists and physiologists.

However, in the 1930s work by two men, K. Z. Lorenz (1938, 1970, 1971) and K. S. Lashley (1938) revived interest in this type of behaviour. Recent work by modern ethologists has largely restricted attention to the relatively "fixed action patterns" or "species-characteristic" type of instinctive behaviour. Today virtually all workers in this field, including even the learning psychologists, admit that the concept of instinct is a valuable one. However much different scientists may disagree about the details of this concept, there is agreement that not all animal behaviour is learned, that many animals are able to display specific and complex behaviour which they have had no opportunity to learn. A review of the work in this area can be found in Thorpe (1963).

How is it possible to distinguish instinctive behaviour from learned behaviour? By performing a simple experiment, such a distinction can be made. If an animal that was removed from its mother at birth and was then kept isolated from all adult members of its own species is then able to display a specific and often complex behaviour which is characteristic of its species, we are then justified in assuming that this behaviour is not learned, that it is instinctive. If such an animal is unable to perform this behaviour then we can assume that it is a learned behaviour.

The purpose of keeping the young animal isolated is to prevent him from observing any behaviour of the adults of his own species, in other words, to prevent the possibility of imitative learning. The isolation also prevents the possibility of the adults communicating any knowledge to him by any type of signals.

Although the young animal should be reared isolated from adult members of his own species, this does not mean that he should be reared in complete isolation. Not only would this be unnecessary but it could very well introduce other complications. Harlow (1965) has shown that young monkeys reared in complete isolation displayed behaviour similar to that which we call neurotic in humans. Obviously a "neurotic" animal that had been reared in complete isolation would not be a good subject for determining if a particular behaviour was learned or innate. In this type of experiment, the young animal should be given as "normal" an environment as possible, with the exception that he never has any contact with adult members of his own species. Several young animals can be reared together, or one animal can be hand-reared by humans or perhaps even reared by other animals of a different species.

Fortunately it is not necessary here to launch into a detailed discussion of all the modern arguments and counter-arguments concerning the concept of instinct. It is sufficient to point out that scientists today recognize that only when a particular behaviour is first performed can it be classified as truly instinctive. Often on subsequent performances the animal is able to show improvement, in other words there is an element of learning involved. For example, although nest-building in wasps and other animals is instinctive, nevertheless individual wasps are able to show some progressive improvement in nest-building ability. The first nest building may be called instinctive, but all subsequent building contains an element of learning. Many instinctive actions are capable of some improvement as a result of experience.

Whenever other animals display behaviour that might suggest evidence of separated foresight, it can be shown that this behaviour is instinctive and not learned. Separated foresight was defined as resulting only from separated learning. Even though many animals display behaviour that is useful for their long term future survival, it does not necessarily mean that this behaviour is the result of learning.

Even though birds fly south in the autumn, this does not necessarily mean they *know* (i.e. have separated foresight) that winter is coming. As will be discussed in more detail below, isolated (hand-reared) birds will still fly south in the autumn. Since the birds have had no opportunity to learn this behaviour from their elders (nor have even experienced cold weather before) this is evidently instinctive behaviour. If a human being had never been taught about (nor experienced) the four seasons and the fact that the weather gets warmer the further south one travels, it is very unlikely that he would head south in the autumn. Presumably a human in this situation might start travelling *after* the cold weather had arrived, but he would just as likely head north or east or west instead of south. Travelling south in the autumn is instinctive for birds, but it is a result of separated learning in humans.

It is not necessary to discuss all of the aspects of instinctive behaviour here. However, one part of Lorenz's theory should be explained. Lorenz has proposed that these instinctive activities can often be seen as "consummatory acts". The pattern of behaviour is apparently the end goal for the animal. The act often completes a chain of reactions and evidently provides a release of tension for the animal. For example Pitt (1931) has shown that the food-burying and hiding behaviour of dogs and foxes is innate. Hand-reared puppies readily perform this activity. The burying itself appears to be consummatory, the dogs display satisfaction even when the food is not in fact adequately hidden and is immediately stolen. Instinctive actions can often be shown to be consummatory acts, completion of the act shows evidence of a release of tension in the animal and is the end goal for the animal.

If instinctive actions are consummatory, then there is good reason for assuming that birds fly south not because they are trying to avoid some future event, but rather because evolution has in some way made flying south itself the end goal. In the same way, dogs evidently do not bury food so that they will have a food reserve tomorrow or next week, rather they bury it because the burying activity itself is tension releasing and therefore probably experienced by the animal as "pleasurable". In contrast, if humans stored away food as a result of separated foresight that food would probably be scarce in the future, the storing itself would *not* be the end goal. The end goal would be consuming the food at some time in the distant future. We would have separated foresight of the long-term benefits of this activity.

With instinctive behaviour, the animal performs the act without knowing the long-term survival value of this activity. He has no foresight about the future results of this activity. In short he does it without knowing why he does it. Evidently he performs this activity only because it is tension releasing to him *now*.

Below is a brief review of some of the better known examples of animal behaviour that might appear to suggest separated foresight. In all these cases, it can be shown that these behaviours are instincitive and not learned. These behaviours reveal only apparent separated foresight in all other animals.

MIGRATION Migration is a long distance mass movement of a species and their subsequent return, usually connected with a change of climate, food supply or breeding season. Although migration occurs in some other animals (such as fishes, reptiles and certain insects), in no other group of animals is it so prevalent and extensive as it is among the

birds. This is probably due to the ability of birds to travel considerable distances in a relatively short time.*

Birds that have been isolated from adult members of their own
species still fly south in the autumn. Therefore this is undoubtedly an
example of instinctive, and not learned behaviour. If birds' migrations
are instincitive, and not a result of separated foresight, we are justified in asking how birds "know" when to start flying and how they are
able to navigate.

Many different factors such as temperature, food supply and even
barometric pressure had been proposed in the past to explain the regularity of birds' migrations. But these factors vary considerably from
year to year which made it difficult to explain the almost calendar
regularity of such migrations. In 1924 William Rowan demonstrated
that it was primarily the seasonal change in the length of daylight that
provided the stimulus for the migratory behaviour.

Apparently birds are affected physiologically by the changing length
of daylight some time before the actual migration. This change is
characterized by an increase of body fat and restlessness. Evidence
that length of daylight is the primary stimulus for the commencement
of migration comes from experimental studies. Birds have been in-
duced to migrate about two months earlier than usual by subjecting
them to artificial daylight periods. Weather also appears to play a
lesser role in starting migration.

Given that the length of daylight is the primary stimulus for birds
to begin their flights, how are they able to navigate? How do they
"know" to fly south? Studies have shown that birds rely primarily on
a form of sun navigation. When the apparent direction of the sun is
altered by mirrors, birds attempt to fly off in the appropriate changed
direction. Those species that migrate at night use navigation by stars.
Also, birds apparently have some sort of "internal clock" to compen-
sate for the movement of the sun or stars during the day or night. The
capacity for navigation is greatly reduced, if not lost entirely, during
heavy cloud cover.

Migration in birds then is an instinctive behaviour that can be accomplished even by birds that have been isolated from birth. Birds
commence their migration in reaction to the changing length of daylight and navigate primarily by the sun or stars. As with some other

---

*Only a few days ago a seven-year-old friend of mine asked me the
following riddle. Question: Why do birds fly south? Answer: Because it is too far to walk!

forms of instinctive behaviour, there is some improvement in performance due to experience. Adult birds are always more accurate naviga-
tors than young birds. Even though learning plays a part in improving
to some extent navigational ability, migration itself is instinctive.
Older birds have no more understanding about why they fly south than
do young birds. The basic "urge" or "drive" to commence flying
south in the autumn is instinctive in all migratory birds. There is no
separated learning involved here, nor separated foresight.

HIBERNATION Although less likely than with migratory behaviour,
some people might claim that animals such as the ground squirrel and
woodchuck hibernate in the winter because they have foresight that
food will be scarce during the coming cold weather. Again even though
this behaviour has survival value for the animals' distant future, this
is not a product of separated learning and foresight. Like migration,
hibernation is an instinctive behaviour.

Although many animals such as bears, skunks, chipmunks and
opossums sleep a lot during the winter, they do not hibernate. Their
body temperature remains high. True hibernation is a state of dor-
mancy in which body temperature is greatly reduced and hence bodily
functions are also reduced. Hibernation is functional for the animal's
future survival because it allows the animal to exist without additional
food during the winter months.

Animals that have been isolated from birth also hibernate. The
primary stimulus for the onset of hibernation is a low temperature.
Other factors such as lack of food and length of daylight also apparently
play a lesser role.

HOARDING Hoarding is the storing or hiding of food in excess of the
animal's immediate needs. Some of the well-known hoarding rodents
are squirrels, beavers, mice and rats. Hoarding by dogs and foxes
has already been mentioned. Hoarding in all these animals can be
shown to have an instinctive basis. Animals that have been isolated
readily perform this activity.

Apparently the hoarding activity itself is the object or the goal of
the animal. Hoarding appears to be a consummatory act in these animals. Rats do not seem to mind their stores being depleted by other
rats, and as already mentioned, dogs do not object even when their
food is immediately stolen.

Most of the experimental work on hoarding has been done with rats.
The primary stimulus for commencement of hoarding is a period of

deprivation. Other contributory factors are low temperatures and the presence of other rats. A deprivation experienced by young rats seems to leave a permanently lowered threshold for hoarding activity in these rats in their later life.

NEST-BUILDING   Nest-building has been studied most extensively in birds, wasps and rats. Nest-building might seem to imply separated foresight because the nest must be built sometime before the young animals are born or the eggs laid. Nest-building appears always to be instinctive, however. This behaviour appears in all normal pregnant rats, even those that have been isolated from birth. Likewise nest-building is innate in birds and wasps.

There is some improvement in building due to experience, however. Although the nest-building motions are innate in birds, they do learn at least to some extent what materials are appropriate for their building. As already mentioned, there is also some improvement in wasps' nest-building due to experience.

Nest-building can be an extremely complicated and elaborate behaviour, especially in birds and wasps. For example, Tinbergen (1953) has shown that the Long-tailed Tit must produce at least eighteen different actions in its nest-building. After a site is chosen, the bird collects moss and places it on the branch, next spider's silk is collected and rubbed on the moss until it sticks and so on. As several scientists have pointed out, it is difficult to account for these involved series of actions without assuming that the bird must have some "conception" of what the completed nest should look like. It seems improbable that this complicated behaviour could be only the result of the completion of one instinctive action triggering the commencement of the next innate behaviour and so on down the line. The bird's behaviour appears to be directional, it certainly seems as if the bird has some "idea" of what his completed nest should look like.

In addition, both birds and wasps are able to temporarily halt construction of their nests and repair damages to the nest which have been artificially produced by humans. These repair activities are quite different from their normal construction behaviour. It is difficult to account for all these varied activities without assuming either a very complicated series of innate movements, or else that the animal must have some innate "image" of what the nest should look like.

Even if these animals do have an instinctive image or conception of the ideal nest, this is not to say these animals have separated foresight. Foresight was defined as resulting only from learning. Since nest-building is instinctive, this "conception" of an ideal nest, even if it exists, must be transmitted biologically. It is not the result of separated learning and therefore can not be classified as separated foresight. Also, even assuming they have such an "image", there is no reason to believe that these animals are aware that the long term function of building a nest is to provide a safe home for their as yet unborn offspring.

FOOD CONSUMPTION All animals including humans consume food. Food provides nutrition for the maintenance of life. Without food an animal would slowly lose weight and strength and finally die. Just because other animals consume food does not mean they are aware of what would happen to them in a few days or weeks if they did not eat. Food consumption is a consummatory act, eating is the end goal of the animal. All other animals consume food only because this is a tension releasing activity for them. In other words, they eat only because they are hungry, not because they know they must eat in order to stay alive. Food consumption is instinctive in the sense that animals reared in isolation will of course perform this activity and that they do not know the long-term survival benefits of this behaviour.

Food consumption is also instinctive in humans. Newborn babies consume milk obviously without knowing why they do so. Some primitive tribes also do not know the long term benefits of food consumption. Malinowski (1929) reported that the Trobriand Islanders were not aware that food was indispensable for life. These natives believed that people eat only because they get hungry.

Of course modern man now knows that consumption of food is necessary for the maintenance of life. This knowledge is the result of separated learning. It takes several days after a person has stopped eating food for the drop in weight and a lack of strength to become noticeable. Before early man could discover this separated relationship, he obviously had to first experience a severe deprivation of food for at least several days.

Even today we usually eat not because of our separated knowledge, but rather because of our instinctive hunger. Like other animals, humans usually eat only because we feel hungry. Only when a person is sick will he be encouraged to eat even when he doesn't feel like it, in order to "keep up his strength".

SEX All bisexual animals copulate. Few people would propose however that other animals are aware of the long term consequences of

this activity. Like food consumption, copulation is an instinctive consummatory act in all animals. All other animals copulate only because
it is tension releasing for them at the present time. No non-human
animal is aware that copulation results in pregnancy and the birth of
young in a few weeks or months time.

Only humans are aware that there is a relationship between sex and
pregnancy. This is a product of separated learning. Several weeks
must pass between sexual intercourse and the first signs of pregnancy.
Human children are not born with this knowledge and they do not know
where babies come from until they are taught the relationship. Several primitive tribes are also not aware of this relationship. This ignorance of the sex-pregnancy relationship will be discussed in much
more detail in the next chapter.

Sexual intercourse is instincitive in all animals. Animals that have
been isolated from birth (but *not* completely isolated) are able to perform this activity. Learning appears to play a role however, especially in the higher mammals. Rats that have been isolated are able
to perform copulation almost "automatically", that is with few mistakes even on their first try. Higher mammals, however, require
some preliminary practice before successful copulation can occur,
especially for males. Male monkeys and chimps that have been isolated are quite awkward and even though obviously aroused, are usually
incapable of successful copulation. Several months or perhaps even
years of practice are necessary before male chimps are proficient in
this activity.

Even though some practice is necessary for perfection of the technique in higher mammals, the basic "urge" or "drive" to copulate
must still be regarded as instinctive. Experienced chimps are no
more aware of why they copulate than are inexperienced chimps. Copulation is an instinctive consummatory act in all animals, experienced
or inexperienced.

Copulation is instinctive in humans as well. As with monkeys and
chimps, some learning of technique is obviously necessary, but it is
reasonable to suppose that young human children that has been reared
in isolation from adults* would eventually copulate and produce offspring. This behaviour would be instincitive in that they would even-

---

*Perhaps in a manner somewhat similar to the children shipwrecked
on an island in the novels *Lord of the Flies* by William Golding or
The Blue Lagoon by H. DeVere Stacpoole.

tually be able to perform the act successfully without instruction from adults and without knowing the long term consequences of this behaviour.

Although all these activities might at first seem to suggest evidence of separated foresight, they can all be shown to be instinctive and not a result of learning. Like the "genetic signals" (discussed in Chapter 2), which sometimes may appear to be similar to our symbolic languages, instinctive behaviours also appear to be similar to the activities we engage in as a result of our separated foresight. However, careful experimentation shows that both genetic signals and instinctive behaviours are different from our activities because they are not learned. Both symbolic languages and separated foresight are a result of learning. Any behaviour in other animals which may at first appear to be similar to our unique abilities can be shown by experimentation to be instinctive or genetically determined.

Although instinctive or genetically determined behaviours have an advantage over our counterpart activities in that it is not necessary for the young animals to learn or be taught these behaviours, this advantage is outweighed by the two disadvantages that these instinctive behaviours are relatively few in number in each species, and that they can be changed to meet new situations only very slowly over many generations by the process of biological evolution.

Chapter 7

Pre-historical Science

Humans are the only animals capable of learning separated relationships. A large part of modern science is engaged in learning additional separated connections. A few examples out of the thousands of known scientific discoveries have already been cited in Chapters 4 and 5. A "scientific discovery" occurs when humans succeed in learning a new separated relationship.

Although science has become a major endeavour in Western culture only in the last few centuries, this is not to say that science did not exist in ancient times. Science has a very long history. Certainly the ancient Greeks and Romans made scientific discoveries. Although the assumptions and philosophy of science were not clearly understood or accepted by the vast majority of people in these early civilizations, there seem to always have been at least a few men who were searching for explanations along the lines of scientific inquiry. A good account of science in early history can be found in Farrington (1969).

Most books concerned with the history of science begin with the Greek civilization, although brief mention is sometimes made of the early civilizations of Mesopotamia and Egypt. However, like many other human endeavours, it is almost certain that science had a pre-historic stage. We have reason to believe that spoken language, art, magic, religion and probably music had a pre-historic stage. It is probable that science also existed before humans began to record events in writing.

Although we will never know the men's names and the exact dates of important pre-historic scientific discoveries, some general information about these early discoveries can be compiled. From archaeological remains it has long been known that early man slowly acquired more and more useful activities and skills such as use of fire, use of metals, domestication of plants and animals and so on. But now with the clear realization that scientific discoveries consist of learning separated relationships, it can be shown for example that when men began farming, this activity was possible only *after* a separated relationship had been learned. In the case of farming, this was possible only after men had learned that seeds placed in the ground result in small plants a few days later. The seed-plant relationship is a separated relationship that only humans are capable of learning. No other animal plants seeds because no other animal has been able to learn the seed-plant relationship. The beginning of agriculture could only occur after a scientific discovery had been made. This seed-plant discovery will be discussed in much more detail later in this chapter.

Many of these early scientific discoveries occurred so long ago and are taught to us at such an early age, that we may find it difficult to realize at first that they must be learned separated relationships. Children are taught that seeds grow into plants at such an early age that it may be difficult for some people to accept that we are not born with this knowledge, and that early men were not yet aware of this relationship. It is conceivable that there might be a few primitive tribes remaining that are still ignorant of the relationship. Certainly this possibility should be carefully explored. Also, a human child might be deliberately kept in ignorance about the seed-plant connection for the first few years of his life and then tested to see if he knew where plants came from.

What criteria will be used to classify the commencement of particular activities by early men as possible only because of scientific discoveries? First, it must be shown that a separated relationship exists. There must be at least two events involved that are separated by time. Second, it must be shown that people who are not taught this relationship are indeed ignorant of it. This must be a learned relationship. This could be accomplished by either locating a primitive tribe that is still ignorant of the particular relationship, or else by testing young children that have not yet been taught this relationship. Third, when possible other animals could be observed and tested to show that all other animals are incapable of learning this relatioship.

Below is a listing and discussion of some of the more important discoveries that pre-historic man made, that must have been *scientific* discoveries. In some cases we have archaeological remains to give us an approximate date when these discoveries must have first been made. In other cases, because of the nature of the relationship, there is no archaeological evidence, and therefore even the approximate date of the original discovery is now a matter of pure speculation.

FIRE
    After the use of crude stone tools, the use of fire was early man's

next great advancement. Present archaeological evidence indicates
that man began to use fire at least 400,000 years ago. Anthropologists
have not found any of today's primitive tribes which do not have the
use of fire. The use of fire is the first clear evidence that early man
was able to learn separated relationships. Before man could begin to
use fire, a scientific discovery was necessary.

Evidence that early men were using fire comes from charcoal found
in caves used by these early humans. It seems clear that men learned
to use fire in two stages. In the first stage, humans were able to "cap-
ture" fire from naturally occurring fires created by lightning or volcanoes and keep the fire going for some time. They learned what to
do in order to keep a fire burning, but were not yet able to make or
create their own fire. During this first stage, man had fire for a time
and then, probably due to heavy rain or human mistake, they suddenly
ceased having fire. The second stage involved creating their own fire.

Humans are the only animals that make use of fire. Other animals
may sit near our fires for warmth, but they have never been able to
reach even the first stage in the use of fire--learning what to do in
order to keep a fire burning.

Early man must have first observed with awe and fear naturally-
occurring forest fires. Attracted by the sight of the red flickering
flames, he must have first felt the warmth and then felt pain as he
tried to reach out and touch this lively warm creature. Frightened
away, he would probably return to the site the next day to find only
smoking grey ashes. What could that strange thing have been? he
must have asked himself. Where did it go? What happened to the
trees that used to be here? Where did all this grey powder come from?

The fact is, it is by no means obvious from observing a wood fire
burning that the fire is consuming or "using up" the wood. The change
in the wood is too slow for our eyes to see. It takes several minutes
after a log is placed in a fire before any change becomes noticeable.
The relationship between wood, fire and ashes is a separated relationship, not a contiguous one.

After observing several fires, one early human must have *hypothesized* that these events must be related in a particular way. After
observing that fire always occurs only "on" wood (and never on earth,
stone or water) and that this wood is always in some way destroyed by
the fire, leaving only grey powder, he must have hypothesized that the
fire "ate" or consumed the wood and that the fire always died after it
had eaten up all the available wood. Like his modern day descendants,
this early scientific genius then proceeded to test his hypothesis by

adding wood to a fire. Sure enough, as long as new wood was added
periodically, the fire continued to burn. The first important scientific discovery had occurred.

It is difficult for us to realize the importance that the use of fire
must have made to early man. Fire provided light at night, warmth
when it was cold, and protection from night predators. With fire early
man could begin to cook his food. Once humans had learned how to
keep this marvellous creature burning, they must have taken every
precaution possible to try to keep their captured friend alive. Of
course we have no knowledge about what actually occurred, but it is
easy to imagine vast stores of wood, various magical and religious
ceremonies to ensure the sacred fire's oontinuation, sentries to ensure
the fire did not die in the night and severe punishment for anyone who
allowed the precious flames to die.

The second stage in the use of fire must have been reached by yet
another scientific giant. After having known fire in his early life, this
person must have searched for ways to re-create fire. One important
relationship that must be learned before fire can be created is what is
now called "kindling temperature". This is the knowledge that very
small twigs and leaves light much more easily than larger branches.
Early man probably first created fire by either producing sparks from
striking flint or by generating heat by friction. After accidentally discovering that flint produces sparks (which certainly appear to be fire-
like), or that friction produces heat (which is also fire-like), this early
scientist hypothesized that perhaps showering these sparks or applying
the heat to a pile of small twigs might re-create his lost friend. With
this hypothesis, man took another giant step towards civilization.

With the ability to create fire, early man could now venture into
colder climates. His movements were no longer restricted to the tropical and sub-tropical regions. Together with the use of animal hides
for clothing, fire allowed the almost hairless Homo sapiens to move
into vast areas of land that had formerly been uninhabitable for him.

ICE-WATER-STEAM

Children are taught at such an early age that ice, water and steam
are related to each other, that it is difficult for adults to realize at
first that this is a learned relationship. It is by no means obvious to
someone who has not been taught this relationship that ice, water and
steam are actually the same material. At least at room temperature,
ice melts so slowly that our eyes cannot perceive the gradual conversion. At normal temperatures, however hard we stare at a block of

ice, we can never actually see the ice changing into water. The change
is so slow that it is beyond the ability of our eyes to detect it.

The knowledge that ice, water and steam are related to each other
is a result of our separated learning ability. After leaving a bowl of
ice standing for several minutes, we return to find part of the ice gone,
and some water in the bottom of the bowl. The idea that the ice slowly
changes into water and vice versa depending on the temperature is
actually only a hypothesis to explain separated events.

There is no archaeological evidence to suggest when humans first
learned this relationship, but there is good reason for assuming that
it was after man learned to use fire. Before the use of fire, men probably seldom ventured far enough out of the tropical and semi-tropical
regions in winter to have contact with snow or ice. Also, before the
use of fire, early men seldom had the opportunity to observe steam.

It is conceivable that some primitive tribes still living in tropical
countries (such as Brazil or the Philippines) have never seen ice and
therefore would still be ignorant of the separated relationship between
ice and water. When first presented with a block of ice, these natives
would have no idea that this is actually only very cold water. If an
anthropologist was careful not to give any hints to the natives concerning the nature of ice and presented them with a new block of ice
each day, the natives would at first learn that the ice always slowly
shrunk and finally disappeared completely. If the ice were placed on
the ground (which is where early man surely found it) even the gradual
appearance of water would be difficult to perceive. Eventually, one
native, after careful observation, experimentation and contemplation,
would hypothesize that the ice did not merely disappear, but rather
always slowly changed into water.

No other animal is aware that ice, water and steam are related.
Animals sometimes eat snow or ice because it immediately quenches
their thirst, but this is not to say they realize that ice is only cold,
solid water. Even though most animals prefer water to ice, no other
animal ever attempts to melt ice (by say placing it in the sun) and then
drink the water later. This proposal could easily be tested in the
laboratory.

BIRTH-DEATH

Although it has long been supposed that only humans know that they
are going to die, until now no one has proposed exactly why this should
be the case. This knowledge is possible only because of our separated
learning. This awareness is actually only a hypothesis that all living

things will eventually die, that death inevitably follows birth.

Again, there is no solid archaeological evidence when this relationship was first realized. People in primitive tribes undoubtedly had a
short life expectancy, life was hard and there were many dangers encountered. Many people probably died either accidentally or from
infections and diseases. The lack of written records of births and
deaths also delayed the realization that all people eventually die.

Even after the birth-death hypothesis was formulated, there was
probably a great deal of reluctance to accept it as valid. The knowledge that death is inevitable is a very difficult fact to accept. There
are still many primitive tribes that believe that death must always have
a definite cause, due to either accident, suicide or sorcery. There
must have been many generations of early men who preferred to believe
that death could be avoided if only they were careful and lucky enough.
This knowledge was undoubtedly the most difficult separated relationship to accept that man had yet encountered.

Even in modern times, people still yearn to believe that there might
be exceptions to the inevitability of death. The Spanish explorers spent
considerable time searching for the "fountain of youth" in Florida.
Today some people have their bodies frozen immediately after death in
hopes that science will some day advance enough to bring them back to
life.

Very young children are not aware that they are going to die. Maria
Nagy (1948) found that children pass through three stages concerning
their belief about death. The first stage lasts until about the age of
five. Children in this stage do not think that death is final. They regard death as only sleep or departure. In the second stage (between
ages five and nine), children accept death as final, but believe that
death can be avoided. With good luck, you can avoid it. The third
stage (from about the age of nine) is essentially the adult view. Death
is final and inevitable. Everyone is going to die no matter how careful
and lucky they are. Early humans may very well have passed through
similiar stages in the acceptance of the death hypothesis.

Virtually all primitive religions and all but one world religion believe in some form of "after-life" (either reincarnation or some type
of eternal spirit life). The varied but almost universal nature of this
belief makes it highly suggestive that this too is an attempt to deny the
hypothesis of necessary death.

Around 70,000 years ago we find early man began burying his dead
with a certain amount of ritual. Skeletons have been found with the
head resting on a stone pillow and the bodies were also provided with

tools and joints of meat. This suggest that man had some type of
belief in after-life. This might also indicate that by this time at least,
he had finally accepted at least the second stage of death--that most
people die, but with luck a few people might escape it. A belief in
after-life probably did not emerge until after early man had finally
accepted the hypothesis of inevitable death at least for most people.
As long as man believed he might live forever, there was no need to
believe in an "after-life". The very term "after-life" implies the
knowledge that life must end.

In all likelihood, eternal life would turn out to be very boring.
Man's unique knowledge that he will someday cease to live is nevertheless very difficult for us to accept. It may be more disconcerting than
death itself. Other animals live in complete ignorance of their eventual death. Death itself usually occurs relatively quickly and is only
sometimes painful. It is the certain knowledge that death is coming
that causes years of intermittent apprehension and dread. Spread out
over many years, the knowledge of mandatory death may well cause
more pain than death itself. Our separated learning ability is not
always a blessing.

FARMING

For an estimated 99% of man's time on earth, he secured his food
by hunting and gathering. He hunted and killed other animals and
gathered fruits, berries, roots and seeds from where he could find
them. Only within the last 10,000 years did humans develop food production. Food production involves the domestication of plants and animals. The first farmers were gardeners or horticulturists, that is
they depended wholly on man power (or more likely woman power) to
cultivate the land. Only when animals had been domesticated could
early man develop true agriculture--the use of animal power to work
the land. The domestication of animals and argiculture will be considered in a later section.

The earliest humans were restricted to only gathering food from
plants because they did not yet know where plants came from. They
did not yet know that either a seed must be placed in the ground before
a plant would grow from that spot, or else a few plants are able to
reproduce by cuttings, that is, a piece of the plant is set into the
ground and this cutting will eventually grow into a new plant. Early
humans had to learn at least one of these separated relationships before they could commence food production.

Sauer (1952) has proposed that early humans learned to create new

plants by cuttings before they learned to crow crops from seeds. Certainly there are several important plants such as bananas, palms,
olives, figs and yams which are capable of vegetative or asexual reproduction which early man could have learned to grow once they had
learned the "cutting" hypothesis. It also seems probable that the cutting hypothesis would have been easier to learn since the cutting is
more imitative or homoeopathic of the plant than is the seed. Also
there are some primitive societies such as some of the Australian aborigines which practise vegetative reproduction (in this case, cutting
off the tops of wild yams and re-burying them), but do not grow plants
from seeds.

But even if humans did learn the cutting method of plant reproduction
first, the number of plants capable of such vegetative reproduction is
quite small. Before farming as we know it (i.e. seed farming) could
commence, humans had to learn the separated seed-plant connection.
The relationship between seeds and plants is by no means obvious, for
example there is certainly nothing about the appearance of an apple
seed for example that resembles the appearance of an apple tree. We
have been taught the relationship between seeds and plants from such
an early age that it is difficult for us to realize that this relationship
is not self-evident. Like the ice melting into water, the growth of a
seed is so slow that it is impossible for our eyes to detect it. Only
with the invention of the motion picture camera in this century and the
use of time-lapse photography can we "see" plants grow. The camera
in effect speeds up time so that our eyes can detect the movement.

The idea that plants come from seeds is actually only a hypothesis
to explain separated events. There are many other possible hypotheses to explain why plants grew here instead of there. Early man in
all likelihood believed in various magical or religious explanations of
the origin of plants before someone proposed the seed-plant or cutting
hypotheses.

In virtually all hunting and gathering societies, the men do the
hunting and the women do the gathering. Also among the societies that
have only simple horticulture, the woman's job is to do the gardening
while the men still do the hunting. Only when farming supplies the
majority of a society's food does it become men's work. It seems quite
likely then that one of the most important scientific discoveries of all
times was made by a woman. As will be discussed later, the commencement of farming initiated a tremendous change in the organization
of structure of human societies.

Seed farming could begin only after one person (probably a woman),

after careful observation and contemplation hypothesized that plants
always grew from seeds. She then tested this hypothesis. At least in
the case of grains such as wheat, early humans undoubtedly ate all the
seeds they could find. There was probably a great deal of scepticism
from the tribe when this early scientist first suggested that the tribe
should take a portion of their valuable food and "throw it away" by
putting it in the ground.

Anthropologists have sometimes suggested that farming began almost accidentally, that gatherers probably dropped some seeds near
their camp and the connection between seeds and plants slowly became
apparent. This could no more be true than the idea that humans gradually became aware of the connection between mosquitoes and yellow
fever. One could just as erroneously say that as yellow fever epidemics only broke out in areas where there were large numbers of mosquitoes, that over the years humans gradually realized the connection
between the two. A hypothesis is always proposed by a single person
at a particular time. Of course, hypotheses are often amended or
changed by other people, but even these amendments are actually new
hypotheses proposed by an individual building upon the work of someone else. Hypotheses do not gradually evolve from no place in particular; they are always the product of one thoughtful individual.

Actually, the fact that most hunting and gathering societies are nomadic, constantly on the move trying to find fresh game and food probably made the discovery of the seed-plant or even the "cutting" relationship more difficult. By the time many of the seeds or cuttings had
sprouted or grown, the tribe had already moved on to a new camp.
Sauer (1952) has suggested that farming began in a society of sedentary
hunters and gatherers, that is, where the supply of food was so abundant (such as a fishing community) that it was not necessary for the
tribe to move about. Considering the difficulties a nomadic tribe would
have had in learning these relationships, this may well have been the
case.

Recent evidence indicates that this discovery was actually made
several times in different societies in different parts of the world.
Often modern scientific discoveries are made at almost the same time
by two or more scientists working independently in different countries.
Apparently once a certain level or stage of culture has been reached,
a particular scientific discovery is ripe for picking. This same phenomenon of independent discoveries may have also occurred in the
discovery of the seed-plant relationship.

With the discovery of these relationships, humans no longer had to

roam about searching for particular plants, for man could now cause
these plants to grow in a chosen place. We could now bring the plants
to us, rather than having to go to them. Man was now able to be a
producer of food rather than just a gatherer.

Also once we had learned how to produce plants, we were then able
to begin selective reproduction of plants. By choosing to grow offspring from only those plants which had desirable characteristics
(from our point of view) we were gradually able to change these plants.
We were able to produce plants with larger yields, with more tolerance
of cold weather and so forth. For example, since bananas are capable
of vegetative reproduction and since the presence of large seeds is an
undesirable characteristic (from the human point of view), through
selective vegetative reproduction, humans have reduced the seeds in
bananas to mere vestigial black specks. The banana plant is now
virtually incapable of seed reproduction, and hence totally dependent
on humans for its survival.

Since the relationship between seeds and plants is a separated relationship, humans are the only animals that are able to learn this connection. We are the only animals that have been able to learn where
plants come from. However, there is at least one other animal that
produces its own food. The harvesting ants and the leaf-cutting ants
also grow their own food. The harvesting ants store their food (usually
grain) in underground chambers. They even bring the seeds out in the
sun to dry if they become damp. It has even been suggested that these
ants "plant" some of the grain outside their nests and then harvest the
crop next year. However, it is very unlikely that this "farming" is
any more than the accidental dropping of grain near the nest.

The leaf-cutting ants do definitely grow their own crop, however.
These ants carry fragments of leaves into their nest and fungi then
grow on the decaying leaves. When a new colony is started, the female
carries a small pellet of fungal spores with her to the new nest. The
fungi produce knob-shaped growths which the ants consume. These
fungi gardens are tended and weeded (freed of moulds) by the ants.
However, it can be shown that this behaviour in ants is not learned.
As discussed in the previous chapter, similar behaviour in different
species may be the result of different mechanisms. The gardening
behaviour in the leaf-cutting ants is an instinctive behaviour. Young
ants which have been isolated will still perform this activity.

Usually gardening or horticulture is a very inefficient method of
food production. The most widely-used tool, the digging stick, allows
only very shallow cultivation of the land. Most horticulturists are

still ignorant of the value of fertilizers for the soil, and as a consequence the land under cultivation is exhausted after a few years and
must be abandoned. Usually horticulturists obtain only part of their
food supply from farming, the rest is obtained from either hunting,
fishing or gathering. Therefore these people are often still migratory,
however their movements are usually much slower than the tribes that
rely on only hunting and gathering.

Even after the discovery of the seed-plant relationship, humans
may not have commenced farming immediately. Studies of modern
hunter-gatherer tribes have shown that as long as their population density remains relatively low, these tribes are often able to obtain ample
food with less work and toil than nearby farming tribes. Thus it is
quite possible that humans discovered the seed-plant relationship
some time before 10,000 years ago, but continued as hunters and gatherers until their increased population forced them to commence farming. Horticulture and especially agriculture produce far higher yields
per area of land than do hunting and gathering.

The main impact of the discovery of the seed-plant relationship on
human civilization had to wait until the commencement of true agriculture. The superiority of farming was not clearly evident until animals
began to be used to till the land. However, some changes were already evident, the practice of horticulture supported a larger population and although still somewhat nomadic, the people could maintain
villages in one location for up to several years. Since gardeners are
able to stay in one place for some time, it allowed these people to
develop more permanent housing, the use of pottery became practicable, and cloth-weaving and cloth garments replaced the garments of
skin used by hunters.

SEX-PREGNANCY

Although today's parents may teach their children the seed-plant
relationship almost without realizing it, they are certainly aware of
their teaching role when they relate to their children the sex-pregnancy
relationship. It is clear that children are born ignorant of this relationship and remain ignorant until they are taught "where babies come
from".

As mentioned in the previous chapter, no other animal has been
able to learn the sex-pregnancy relationship because this is a separated relationship. Sex in all animals is instinctive; it occurs in animals that have been isolated from birth. Although other animals
copulate, they do so only because it is instinctive and tension-reducing

for them at that time, they are not aware of the long term consequences
of this behaviour. Early humans were also ignorant of this relationship. Like all other animals, early humans engaged in sexual activities only because it was instinctive and pleasurable.

This discovery of the sex-pregnancy relationship was a scientific
discovery. There is little evidence available at present to show when
this discovery was first made. Anthropologists have found several
primitive tribes that were either completely ignorant of this relationship or else had only a rudimentary understanding of this connection.

Although modern scientists consider these tribes ignorant of this
relationship, the natives themselves would not have accepted that they
did not know where babies came from. As discussed in Chapter 5,
primitive peoples are able to hypothesize and theorize just as well as
we are. People from primitive tribes usually have an explanation for
all events that are of great importance to them, but many of their explanations and theories are of a magical or religious nature. The
road from a completely magical, superstitious hypothesis to a scientific hypothesis of the same phenomena is a long and twisting journey.
There is often at least an element of empirical validity in even the
most magical or superstitious of theories.

Perhaps the most vivid example of a primitive tribe's ignorance of
the sex-pregnancy relationship was recorded by Bronislaw Malinowski
in his classic book, *The Sexual Life of Savages* (1929) which describes
the behaviour of the people of the Trobriand Islands off the north-east
coast of New Guinea.

Although these natives were ignorant of the sex-pregnancy relationship, they nevertheless had a quite complex theory explaining where
babies came from. The theory was also bound up with their ideas
about reincarnation and spiritual beings. They believed that after
death, a person's spirit (baloma) leaves the body and moves to the island of Tuma--the island of the dead. The spirit lived on this island
for some time but eventually tired of this existence and longed to return. These spirits had an amazing ability to jump back in age and
become a small pre-born infant. The natives believed that the only
way a woman could become pregnant was when one of these rejuvenated
spirits found its way back to the Trobriands and into the womb of the
woman.

There was some difference of opinion among the natives as to exactly how these spirits returned to the Trobriands. One version was
that the spirits returned by floating in the sea and then drifting around
the shores waiting for a chance to enter the body of a woman while she

bathed.

The natives were completely ignorant of any connection between sexual activity and pregnancy. For example, they had no knowledge of the physiological function of the testes. They believed the testes were only an ornamental appendage, "How ugly would a penis look without the testes" (p. 168) the natives would say. Also, a man whose wife had become pregnant during his prolonged absence would happily accept the child, and he would see no reason to suspect her of adultery.

The children had almost complete freedom to learn and indulge in sexual activites as they wished. They initiated each other in sexual behaviour long before they were actually able to carry out the act of sex. The adult's attitude towards this infantile activity was either complete indifference or else complacency. Children began their real sex life at about the age of six to eight for girls and ten to twelve for boys.

This belief meant of course that the natives were not aware of the father's role in reproduction. They believed that the child came exclusively from the mother. For them the term "father" had only a social definition, that is, the man married to the mother. The whole process of reproduction lay with the spirits and the females. The men had nothing to do with it. This belief also probably explained why they were a matrilineal society, all descent, kinship and social relationships being accounted only in terms of the mother.

However the Trobrianders did accept one small fact that showed that even their theory was based at least in part on empirical observation. They were fully aware that a virgin cannot conceive. However, their explanation of this differed somewhat from ours. They believed that this was because the vagina must be open before a spirit could enter the womb. The vagina of a virgin will not allow the spirit to enter. After the physiological obstacle had been removed (either through sexual intercourse or in some other manner), the woman is able to become pregnant. With the removal of the maidenhead, the woman can conceive without ever having sexual intercourse. Malinowski was somewhat amazed that they had even reached this first level in the understanding of pregnancy. Since there were no virgins in any of the villages, it is difficult to see how they learned even this relationship.

The natives were fully convinced that a girl could become pregnant without previous sexual intercourse. Malinowski described some of the arguments he had with the natives trying to convince them that they were wrong. These arguments are quite illuminating because they

show clearly that even though their theory had a large component of
religion and superstition, they nevertheless sometimes used empirical
evidence to back up their assertions. Malinowski was unable to convince them that their theory was incorrect. In some respects the arguments resemble the arguments of modern day scientists of different
schools who hold different theories about a certain phenomenon. Each
side emphasized the evidence that supported their particular theory.

For example, the natives pointed out that some women were so ugly
and repulsive that no one believed that they could have ever had intercourse (except of course for the few men who knew better, but were
too ashamed to admit it). The natives repeatedly pointed out that even
though these ugly women could never have had intercourse, they nevertheless had children. The natives would ask Malinowski, "Why did
they become pregnant? Is it because they copulate at night time? or
because a baloma has given them children?" (p. 185).

Also they would bring up the many cases of unmarried girls who had
plenty of intercourse, but had no children. The unmarried girls had
a much more active sex life than the older married women, but few of
the girls had children. "Unmarried girls continually have intercourse",
the natives would say, "in fact they overflow with seminal fluid, and
yet they have no children" (p. 188).

Of course, a few of the unmarried girls had children, but not nearly
as many as one would have expected. Malinowski was not able to explain this adequately, except to say that the young girls did indeed
seem to be sterile during their period of licence. Today's scientists
now refer to this phenomenon as "adolescent sterility". Apparently
most girls remain sterile for a year or two after they first begin to
menstruate. Since the girls in the Trobriands began their sex life at
about the age of six to eight, they had several years of sexual intercourse before they became fertile. Many of the girls were already
married by the time they became fertile.

The Trobriander's theory of pregnancy did not recognize any connection between sex and pregnancy. However, on this one small point,
their theory recognized an aspect of human sexuality that scientists of
Malinowski's time were not yet aware of. The Trobrianders were
aware that young girls who had sexual intercourse seldom became
pregnant. On this one small point, their theory was more "advanced"
than was the Western scientific theory of the day. Because of social
prohibitions in Western cultures, young girls never have an active sex
life, and consequently we had no way of learning about "adolescent sterility" until scientists visited other cultures where this practice was

allowed. The Trobrianders were aware of one tiny bit of knowledge
of which we were still ignorant. Even very primitive theories often
are based at least in part on empirical observation.

Anthropologists have studied several other societies in which the
sex-pregnancy relationship is not clearly understood. Although some
of these cultures' theories are a little more advanced than the Trobriander's ideas, in that they recognized some sort of vague connection
between sex and pregnancy, they are still seriously deficient from our
point of view. For example, Marshall & Suggs (1971) reported that
although the people of Mangaia, a small island in the South Pacific,
understand that sex is related to childbirth, they believe that pregnancy
results only from having intercourse repeatedly with one man. Thus,
parents encourage their young daughters to sleep with many different
men, so that they will not become pregnant until they find the right
man.

Probably because many primitive tribes have some vague realization of the sex-pregnancy relationship (even the Trobrianders accepted
the virgins could not become pregnant), some anthropologists have
doubted whether these tribes were truly ignorant of this connection.
They have proposed, for example, that the anthropologists who studied
these tribes must have misunderstood what the natives really believed.
Or perhaps that the natives deliberately deceived the anthropologists
or perhaps even that the natives had deceived themselves into thinking
they were ignorant of this connection! This has led to a tremendous
and prolonged debate within the field of anthropology. The most recent
proponent of this outlook is Leach (1966). However, most anthropologists today would agree with Spiro (1968) and Montagu (1974) who
concluded that at least some of these tribes were genuinely ignorant of
the sex-pregnancy relationship, and that even in the tribes which
showed some limited awareness of the connection, this was probably
preceded by a stage in which there was no awareness of the role sexual
intercourse played in causing pregnancy.

It is not known when humans first hypothesized a relationship between sex and pregnancy. Even after it was first proposed, there
were many more details of the process of conception to be learned.
The detailed knowledge of fertilization of modern science has been
acquired only recently. For example, Descartes (1680) in a posthu-
mously-published paper, expressed the view that the process of sexual
fertilization resembled chemically the process of beer-brewing. The
froth of beer can be used as yeast for other beers, likewise, Descartes
believed that the sexual fluids of both sexes mixed together and acted

as yeast on each other to commence fertilization.

Spermatozoa were not seen until 1677 when a student named Ham put some semen under a microscope. It was exactly 200 years later, in 1877, that the entry of a sperm into the ovum (of a starfish) was first observed by Fiol.

The discovery of the sex-pregnancy relationship had far reaching consequences on human civilization. These consequences will be discussed in more detail in Part Three, however it should be clear here that the first result of this discovery was that humans now had an expanded ability to control their population. Before the discovery of the sex-pregnancy connection, only two methods of birth control were available, abortion and infanticide. For humans who did not know what caused women to become pregnant, the only methods of limiting population were to abort the foetus or else kill the newborn child. With the knowledge of the sex-pregnancy relationship, however, humans could begin to use other methods of birth control. Admittedly, for primitive peoples who had only a vague knowledge of the sex-pregnancy relationship, the only additional methods of birth-control which this allowed were abstinence and perhaps coitus interruptus. With the additional knowledge of the function of the sexual organs, sterilization became possible. Ovariectomy was carried out by some Australian tribes and also in ancient Egypt for the purpose of sterilization. Castration might also have been used for contraceptive purposes.

When it was first recognized that there were certain periods when women were infertile, the practice of limiting sexual intercourse to the "safe period" could have begun. Early humans may also have practised placing a barrier in the vagina before intercourse to try to prevent conception. A suppository made from crocodile dung was used in Egypt over 3,000 years ago. The dung provided a plug and the acid in it acted as a spermicide.

DOMESTICATION OF ANIMALS

It is not clear from the evidence now available whether humans domesticated plants or animals first. Both may have occurred at more or less the same time in different areas. Dogs were in all likelihood the first animals to be domesticated, at least 10,000 years ago. The domestication of the dog was almost certainly carried out by hunters. Hunters often keep the young of various animals as pets.

Although humans must have learned the seed-plant relationship before they could become farmers, it was not essential for men to learn the sex-pregnancy relationship before they could have domesticated

animals. Animals breed instinctively, and as long as man did not
interfere with their breeding habits too much, the animals would reproduce on their own. As long as early man only used domesticated
animals for pets or as an occasional source of food, it was not necessary for him to have learned the sex-pregnancy relationship; however
it is doubtful if man could have commenced domestication of animals
on a large scale until he had learned this relationship.

For example, the Trobrianders were not aware of the sex-pregnancy relationship, but nevertheless had domesticated pigs. When
Malinowski asked how pigs reproduced, the answer was that "the female pig breeds by itself" (p. 190). By this they meant that there was
no spirit involved in reproduction as in humans, the female pig reproduced on her own.

The Trobrianders made a big distinction between their tame village
pigs and the wild or bush pigs. The village pig was considered a great
delicacy, while they had a strong taboo against eating the bush pigs.
However, the natives allowed their female village pigs to wander to the
outskirts of the village where they copulated with the male bush pigs.
Also, in order to improve the condition of their male village pigs, the
natives always castrated them. This meant of course that all the offspring were actually descended from the wild pigs, but the natives
were completely ignorant of this.

The natives firmly believed that if a female animal was isolated
completely from all males of the species, this would in no way interfere with her ability to reproduce. They even used the reproduction
of pigs to back up their argument that there was no connection between
sex and pregnancy. "From all male pigs we cut off the testes. They
copulate not. Yet the females bring forth" (p. 191). They conveniently
ignored the activities of the wild pigs.

Without the knowledge of the sex-pregnancy relationship, the domestication of animals was always on a precarious foundation. If it were
not for the wild male pigs, the Trobrianders would have unknowingly
exterminated their domesticated pigs. This unfortunate event probably
occurred many times to early humans. Since male members of most
species are usually stronger and more difficult to control and tame,
there would be a strong tendency for humans either to castrate all the
males or else kill them before they reached full maturity. After this
had been done, the female animals would have mysteriously ceased to
reproduce.

As long as early man only used domesticated animals as pets or
perhaps had them occasionally as a source of food, he would probably

not have interfered very much with their breeding habits. But as soon as he began to try to domesticate them on a large scale and use them for one of his major sources of food, he would have probably begun interfering with their reproduction. Until man had learned the sex-pregnancy relationship, all such attempts at domestication were probably doomed to eventual failure.

Whether humans first learned the sex-pregnancy relationship concerning animals and then generalized the same relationship to humans or whether it was first learned with humans and then generalized to animals is not known. Since the discovery of the sex-pregnancy relationship may well have occurred more than once in different societies, the order of the discovery need not have been the same in each society. However, as modern scientists have found, it is usually much easier to run controlled experiments with animals than with humans. Likewise, a pre-historical scientist might have found it easier to test his hypotheses with animals rather than with humans. For example, he would have probably found it easier to isolate and prevent a female pig from copulating than to do the same with a human female.

The essence of the domestication of animals lay in causing them to breed successfully while dependent upon humans. Only after man had learned the sex-pregnancy relationship could he gain the full benefits of domesticating animals. The most important such benefit was that man could begin using animal power in his farming. Horticulture could become agriculture. With draft animals such as horses, cattle or donkeys and a plough humans could cultivate the soil much more efficiently than could be done with a digging stick. Heavier soils could be cultivated and much larger areas could be planted at one time. True agriculture was confined to the Old World, probably because none of the native American animals were suitable as draft animals except the dog, which was used only in the Arctic to pull sledges.

Also, after man became capable of domesticating animals on a large scale it was possible for him to keep large herds of animals such as sheep, cattle, horses, goats, reindeer and camels to supply most of his food. People that obtain most of their food in this way are usually referred to as "herders" or pastoral peoples. Pastoral peoples are usually nomadic, moving their herds to new pastures at certain times of the year. All pastoral peoples were found only in the Old World because of the very small number of suitable animals amenable to domestication found in the New World.

The domestication of animals proved to be a significant advance in human culture, primarily because it allowed true agriculture to begin.

We have seen that before agriculture could begin humans had first to
learn both the seed-plant relationship and also probably the sex-pregnancy relationship. The use of domesticated animals in connection
with farming may also have led to yet another scientific discovery, the
use of fertilizers. Nearly all horticulturists are ignorant of the process of fertilizing the land. The first farmers found that after farming
a plot of land for a few years, the crops steadily declined. The value
of manure as a fertilizer might have first been hypothesized by a
farmer who noticed that fields which had been grazed over by his domesticated animals grew better crops than did those fields that had not
been grazed. Early humans also learned to use other materials such
as dried blood, guano, fish, ground bones and wood ash as fertilizers.
With the use of fertilizers humans could cultivate one plot of land
almost indefinitely. This meant of course that man no longer had to
move to new land every few years.

With the start of agriculture, many other changes in human civilization could begin. Since agriculture supplied a much larger amount
of food, population grew enormously in agricultural societies. Villages became more stable and trading between different societies
increased. Lifelong specialization of occupation could increase, which
led to many new inventions such as dairying, the wheel, writing and
metallurgy.

All these advances depended upon an agricultural base of food production. Agriculture provided the stability and abundant food necessary to free humans from their constant search for food. Most of the
great historic urban civilizations such as Egypt and Mesopotamia were
built on an agricultural base. Although some complex cultures such
as the Aztec and Inca cultures in the New World developed on only a
horticultural base, these cultures probably would have developed much
farther if the New World had had more draft animals suitable for
domestication.

It is arguable then that the discovery of the seed-plant relationship
and of the sex-pregnancy relationship were the most important scientific discoveries of all times. It was only after these two discoveries
had been made that humans could cease being nomadic and develop
true agriculture. All other advances in human civilization depend on
a stable society and high yields of food which true agriculture supplies.
It is unfortunate that we will never know the names and circumstances
of the persons who made these most important of all scientific discoveries.

POTTERY

Hunters and gatherers invariably use basket or hide containers because of their strength and flexibility. The use of pottery containers probably did not begin until man commenced farming. Pottery is too fragile to be used extensively by nomadic peoples. Although pottery might have been discovered before humans learned the seed-plant relationship and began farming, it was not a practical material until that time.

The production of pottery is the result of yet another scientific discovery. The relationship between clay, heat and pottery is a separated relationship which no other animal is capable of learning. Anthropologists have often suggested that man might have "accidentally" discovered pottery after burning a basket which had been plastered with clay to make it water-tight. This may very well have occurred, but even if so, the discovery was no more "accidental" than many other scientific discoveries. The burning of a basket plastered with clay probably occurred many times in our pre-history before man discovered pottery. And probably many times someone noticed small lumps of hard stone in the ashes of the fire the next morning, but even after both events had occurred, this did not ensure the discovery of pottery-making. The production of pottery could not have begun until one person, while rubbing the lump of hard brittle stone in his hand, asked himself "Where did this come from?" Pottery making could not begin until one person *hypothesized* a connection between clay, heat and this new material. It is the ability to hypothesize that lies behind every scientific discovery and marks us apart from all other animals.

A great deal of additional hypothesizing and experimentation must have gone on before man could produce pottery of strength and durability. Natural clays vary in consistency, so in order to produce suitable pottery, other materials such as sand, mica and lime must often be mixed with the clay to make it more or less plastic or sticky. The clay must be shaped so that internal stresses will not cause cracks, it must be properly dried before firing and just the right amount of heat must be applied to insure proper conversion to pottery.

Untreated pottery is somewhat porous and allows liquids to slowly escape. However, with the discovery of glass, man learned to apply a coat of material similar to glass to his pottery and produce a smooth, waterproof finish. We now call this technique "glazing". The Chinese, after centuries of experimentation, finally were able to learn how to mix glaze with the clay and produce porcelain or "china".

USE OF METAL

The technique of smelting was a scientific discovery. Smelting
involves freeing metal by heating the ore. Somewhat similiar to the
production of pottery, smelting involves a separated relationship between stone, heat and metal. The ores of copper, tin or iron do not
look at all like the metals which can be extracted from them. Even
after early man had repeatedly found small globules of metal in the
ashes of his dying fire, production of metal could not begin until someone hypothesized that only when the fire was built on a certain type of
stone did these beads of metal mysteriously appear.

Early humans had a very limited use of metal before the discovery
of smelting. Copper occasionally occurs in a free (relatively pure)
form and required no smelting to make it usable. This was sometimes
cold-hammered into ornaments and tools by early man.

Even after the discovery of smelting, metal tools were no real
threat to stone tools. Stone tools are superior to tools made of copper
or gold. In their pure form these metals are too soft to make useful
cutting tools. Humans did not begin to use metal tools on a large scale
until someone discovered how to make alloys--the mixing of different
metals in particular proportions. The first alloy to be produced was
bronze. Bronze produced by early man is about 90% copper hardened
by small amounts of either tin, gold, silver, arsenic or phosphorous.
Although the first production of bronze might have been accidental due
to smelting copper ore naturally mixed with other ores, the conscious
production of bronze could not have commenced until a pre-historic
scientist proposed that a mixture of different ores in a certain proportion would produce a harder metal than the metal resulting from
only one ore.

The techniques of smelting and alloying did not occur until about
5,500 years ago somewhere in the Near East. It was about this same
time that man began to record events in writing. The discovery of
smelting and alloying were the last major pre-historic scientific
discoveries. With the advent of writing, historical science could now
begin.

Chapter 8

Magic, Religion & Science

There is no great scarcity of people who think they know
something but really know little or nothing . . . In this one
small point then I seem to have the advantage . . . I am wiser
than they are only in that they know nothing, and think that they
know; but I neither know nor think that I know.

Socrates

When did the first human exist? When did our species evolve?
Anthropologists have been studying this question for many years and
have made many remarkable discoveries, but there is still a great
deal of uncertainty concerning the origin of mankind. There are still
too many unknowns involved to give a date of our origin. To some extent it is still an arbitrary decision as to where we draw the line and
call this early ancestor's skull "human".

The earliest man-like creature existed at least two million and
perhaps as much as three or four million years ago. This hominid had
bipedal progression, made very simple stone tools and had a brain
capacity of about 650cc to 800cc. This compares with 550cc for the
gorilla and 400cc for the chimpanzee. About one million years ago
Homo erectus evolved. This early man had more complex stone tools,
is generally assumed to have used fire and symbolic language. His
brain capacity averaged around 900cc. Only half a million years ago,
Homo sapiens arrived. Our brain capacity is about 1450cc.

There is still a great deal of disagreement among anthropologists
about the details of this broad outline, and the dates are subject to
change from year to year as new specimens are found. Fortunately
the details of the evolution of mankind are not too important here. The
broad framework is clear--humans evolved from earlier hominids--
at least two million years ago early man began using very crude stone
tools and at least 400,000 years ago he began to use fire.

Since the thesis of this book is that humans have a unique type of
intelligence, namely that only humans are able to symbolize and hy-

pothesize, it will be sufficient for our purposes to assume that earlier hominids became human when our ancestors were first able to symbolize and hypothesize. The evolution from non-human to human in all likelihood was the result of one or more chance mutations which resulted in the increased size and function of our brain. For reasons which will be explained in the next chapter, it will be assumed that humans acquired these two unique abilities at the same time.

Although we do not know exactly when this transformation took place, it must have been at least 400,000 years ago when present anthropological evidence indicates that early man began using fire and may very well have occurred much earlier. Although chimps are able to make simple tools from sticks and leaves, as was discussed in Chapter 1, they have not yet been able to use one tool to make another tool. The making of all but the crudest of stone tools by early man would have required precisely this ability. It may very well be that using one tool to make another tool requires a hypothesis--there would certainly be a considerable time lag between the finding or production of the first tool and the use of the second tool. If this is the case, humans may well have evolved over a million years ago, since our ancestors began using more advanced stone tools about this time.

Once our ancestors had acquired the ability to symbolize and hypothesize, they could then develop symbolic language and begin attempting to learn specific separated relations. We know this happened at least 400,000 years ago, if not much earlier. Once they had acquired these two unique abilities, it can safely be assumed that these early humans were on an intellectual par with modern man. At least 500,000 years ago early man had the same brain size as modern man.

If early humans were as intelligent as we are, how are we to explain the fact that early scientific discoveries came so few and far between? Why did early man have such a hard time learning separated relationships? Hundreds and even thousands of years often passed between major scientific discoveries in the pre-historic world. For example, even if the first humans evolved only 500,000 years ago, why did it then take us 490,000 years to discover even the simple seed-plant separated relationship? Even assuming this very recent date of human evolution, this would have meant that man waited 98% of his time on earth before he discovered the seed-plant connection. If early man was our intellectual equal, how do we explain his scarcity of scientific discoveries compared with our abundance? What has modern man acquired in recent centuries that early man did not have?

Science is based on man's unique ability to hypothesize. However,

as mentioned in Chapter 5, magic and religion are also based on the
ability to hypothesize. Science, magic and religion are all dependent
on our ability to propose relationships between events separated by
time. The people in today's primitive societies show that they are
quite capable of hypothesizing. However their hypotheses are mostly
of a magical or religious nature, rather than of a scientific nature.
For example, it is a Tasaday belief (Nance, 1975) that if a particular
type of bird sings or appears when they are preparing to go somewhere,
this is a warning of danger and consequently they delay their trip.
This belief is actually a hypothesis which proposes a connection between two events separated by time.

Science is distinguished from its two cousins, magic and religion,
by its requirement that all hypotheses considered should be testable,
or in Popper's (1934) terminology, falsifiable--that is it should be
possible at least in principle to prove the hypothesis wrong! People
of today's primitive cultures are just as able to propose hypotheses as
are the people in modern cultures, but the idea of only considering
hypotheses which are testable and falsifiable is a relatively modern
requirement. Like the people in today's primitive societies, early
man must have been able to hypothesize, but his criterion for the
selection and rejection of his hypotheses differed from that of modern
man.

If early man had a scarcity of scientific discoveries, he also must
have had an abundance of magical and religious beliefs. Early man's
intelligence surely led him to hypothesize about the many strange
events that surrounded him, but because he had not yet acquired the
notion of carefully testing his hypotheses before acceptance, his scientific discoveries were rare.

The requirement that hypotheses should be testable and falsifiable
is so prevalent in our modern scientific world, that it is sometimes
difficult for modern man to understand why early humans had such a
hard time recognizing this fundamental notion. The idea of empirical
testing seems so "natural" and obvious to most of us, it is difficult
to imagine why primitive man did not see it sooner. Since all humans,
and also other animals, use the "trial and error" method in contiguous
learning, it seems to us a simple step to use a similar experimental
method in separated learning. But as archaeological evidence has
shown, empirical science had a very slow beginning and today's primitive peoples have other ideas about which hypotheses they choose to
believe in. Early societies probably held beliefs that were primarily
magical and religious.

This classification of beliefs into magic, religion and science was first proposed by Tylor (1891) and Frazer (1890). Since that time numerous objections have been raised to their proposals. Many of these ideas have been completely abandoned by anthropologists and most other parts of their theories are now accepted only in a modified form. In fact many of today's anthropologists would prefer to drop this system of classification completely. However, for purposes of explanation, I will first present a highly modified and sophisticated version of this system of classification and then later in the chapter I will be able to explain why some anthropologists would prefer to abandon this system altogether. Magical beliefs will be discussed first.

MAGIC

There has never been a primitive society found that did not have magic and religion. It is also important to note that all primitive societies have at least some scientific knowledge. It is probably safe to say that all three forms of belief have been present to some extent in all cultures. What varies from one culture to another is the extent and the frequency of each of these three belief systems. A very readable summary of the work done in the anthropology of magic and religion can be found in Howells (1948).

To many people today the word "magic" conjures up images of a man on a stage pulling rabbits out of a top hat. This is only a vestigial meaning of the word, however. These sleight of hand tricks should not properly be called "magic". Properly speaking, magic means the formulas and spells which humans use in an attempt to control their environment. For example, a tribe in western British New Guinea believes that if a man who has killed a snake, burns it and then smears his legs with the ashes, then no other snake will bite him for at least several days. This of course is a hypothesis that snake ashes act as a sort of snake repellent.

Unlike religion, magic is impersonal. Magic makes no appeal to gods or spirits. Anthropologists have sometimes referred to magic as a pseudo-science. The magician believes that by properly performing a certain act, he can manipulate nature. With the correct rite or spell, it is believed that man can directly control the weather, animals and crops. Like science, magic makes hypotheses proposing connections between separated events.

A *magical* belief will be defined here as a hypothesis which proposes a separated connection which *present day scientists* do not ac-

cept as valid. It is important to note the rather tenuous and temporary
nature of this definition. Although we believe *our* scientific hypotheses
to be correct and valid, in a few years time future scientists will undoubtedly reject some of today's "scientific" ideas. Likewise, we now
believe certain hypotheses proposed by primitive tribes to be invalid
or magical, but this designation could easily change in the future. We
might some day accept that certain primitive beliefs which are now
labelled "magic" were actually "scientific". Some anthropologists
may consider this definition of magic very broad. Often anthropologists make distinctions between "magic", "superstition", "folk-medi-
cine" and even "ethnoscience". However, I am defining magic in the
broad sense to cover all these terms.

Magical beliefs abound in primitive societies. However, we are the
ones that classify these beliefs as "magical". To the primitive man,
all his beliefs are "scientific", that is, he believes all his proposals
are valid and have been supported by experience. His requirement of
empirical testing may be much less rigorous than ours, but he always
believes that his ideas have been verified by experience.

For example, early man learned to use manure as a plant fertilizer.
We accept this to be a scientific hypothesis. Likewise, the New England Indians taught the Pilgrims to put a fish in each hill of corn to
make the corn grow better. We also accept this as a valid hypothesis.
However, when planting grain, a man from the Zande tribe will put into
the ground the juice from one of his medicine plants. We know that the
juice from this plant is not effective and therefore call this practice
"magic". But the Zande believes that his juice is useful just as the
New England Indians believed that fish were effective   Magic is always
"scientific" to the man who is using it.

Thus there is a very close relationship between magic and science.
Also, remember that before Semmelweis, the "scientific" explanation
of childbed fever was the "atmospheric-cosmic-telluric" epidemic influence. Scientists of the day would have used what they regarded as
empirical evidence to support this theory. However, present day evidence would not support this idea. A hypothesis which was once regarded as scientific is now classified as invalid. Scientists usually
refer to these cast off hypotheses as "outdated science". However, if
an anthropologist from a highly advanced civilization had visited Europe
during this period he might well have labelled this idea as "magical"!

People in primitive tribes believe their hypotheses are scientific,
and Semmelweis' colleagues believed that their proposal was scientific.
Today we reject both of these assertions. We label the primitive man's

beliefs as "magical" and the "atmospheric-cosmic-telluric" influence
as "outdated science". But in both cases, these beliefs were scientific
to the men using them and in both cases we currently do not accept
them as valid. Beliefs which *we* label "scientific" may also be reclas-
sified some day.

In other words, humans in different societies and even in the same
society at different times, have different ideas about which hypotheses
are in fact valid. Our labelling system of "scientific", "outdated science" and "magic" reflect not only these differences, but also imply
that what we believe today is "right", but what our grandfathers used
to believe or what other societies believe is "wrong". But as discussed
in Chapter 5, since we can never be certain that the hypotheses we
accept are, in fact, valid, the logical consequence of this is that often
we cannot be certain that other people's hypotheses are, in fact, invalid.

This is not to suggest that magic and science are completely identical. On the contrary, as we shall see, once the hypotheses have been
proposed, magic and science deal with them in somewhat different ways.
An excellent analysis of the similarities and differences between magic-
religion and science can be found in Horton (1967). I am only proposing
that magic, outdated science and science are identical in that they all
attempt explanations of the world by proposing hypotheses between
events separated by time. They all spring from our unique ability to
"hypothesize".

On what basis does primitive man propose his hypotheses? Are
these hypotheses proposed in a strictly random manner, or are there
any similarities which most magical hypotheses seem to have in common? James Frazer in his famous compendium of primitive magic
and religion *The Golden Bough* (1890) found that most magical beliefs
fall into two categories. He called these two principles the Law of
Similarity and the Law of Contagion.

In the first of these categories, the Law of Similarity, the hypotheses are based on the assumption that "like produces like" or that an
effect resembles its cause. The magician believes that he can produce
an effect merely by imitating it. Probably the best known example of
this principle is when a primitive magician attempts to injure or kill
an enemy by destroying an image of that enemy, the hypothesis being
that as the image suffers, so must the man.

This type of magic is not only used for injuring enemies, but also,
for example, for helping cure the sick. This is sometimes referred
to as homoeopathic magic. The ancient Hindu's cure for yellow jaundice was to dab the patient with yellow porridge and then to wash off

the porridge (and supposedly the jandice as well). Then to give the
patient a healthy red complexion, some hairs of a red bull were glued
to his skin.

There is to this day a controversial branch of medicine known as
homoeopathic medicine. Medicine is prescribed on the basis that
"like cures like". This is based on the assumption that a drug which
produces disease symptoms in a healthy person would cure a sick
person with similar symptoms. This is not to say that this branch of
medicine is necessarily invalid. It is possible that some or even all
of these medicines are effective. But to be classified as *scientific*
each of these prescriptions must be shown by careful empirical testing
to be effective.

Frazer maintained that primitive taboos are only a negative appli-
cation of the law of similarity. A taboo is a prohibition against a certain activity. The taboo is always based on a hypothesis that certain
undesirable consequences will follow this activity. For example,
Eskimo boys are forbidden to play cat's cradle in the belief that if they
did so their fingers might similarily later become entangled in the har-
poon-line. When a magician engages in a certain activity which he
believes will have a beneficial consequence we call this a magical spell
or charm. When he refrains from another activity which he believes
will have an undesirable result, we call this a taboo. Both behaviours
are based on hypotheses which he believes to be valid, but which we
do not believe are supported by empirical evidence.

The second principle of magic is known as the Law of Contagion.
In this case hypotheses are based on the assumption that things which
have once been conjoined must remain ever afterward, that whatever
is done to the one must similarly effect the other. This principle of
contagion is interesting because it in effect assumes that events which
formerly had a contiguous connection, must still have some relationship with each other, even though they are now separated in time.

A familiar example of this type of magic is the idea that a connection is supposed to exist between a man and any severed portion of
this person such as his hair, teeth or nails. A magician who obtains
any of these objects from a person is believed capable of working his
will on that person. This principle is also supposed to hold between
a wounded man and the agent that caused the wound. For example, in
Melanesia, if a man's friends or family can find the arrow which
wounded him, they will keep it in a damp place or in cool leaves as it
is believed that this will cause the inflammation of the wound to subside.

These are just a few examples out of the literally hundreds of cases

cited by Frazer. Although most magical beliefs fall into these two
principles or laws, this is not to say that the primitive man is consciously aware that he is using these two principles to propose his
hypotheses. This is almost certainly not the case. When faced with
the totally unknown world of mysterious events, literally millions of
hypotheses are possible to explain any one event. When confronted
with all these uncountable possibilities, primitive man must have unconsciously sought some guidelines to narrow the range of possibilities.
In some cases these guidelines may have actually worked. Remember
that in discovering how to create fire, some early human must have
hypothesized that placing sparks from a flint or heat from friction
(both of which are fire-like) near small twigs or leaves might re-
create fire. Either of these hypotheses would have been based on the
assumption that "like produces like".

When first proposed, these hypotheses seemed just as reasonable
and likely to be valid as any other, if not more so because they had
some sort of identifiable connection with the consequence--they either
resembled the consequence or else had once been in contact with it.
Blame should not be placed on the people who first proposed these
hypotheses, but rather on their followers who continued to accept these
explanations for generation after generation. Early man certainly had
as much if not more need to learn about his surroundings as we do.
Famine, disease and danger threatened him every day. Rather than
condemning early man for his attempts, we should admire him. The
pre-historic man had nothing to build upon, no guidelines to help him
learn separated connections. If these early attempts were feeble
compared with ours, it is only because he faced the unknown first. It
is his meagre beginnings that we build on today.

The desire to foretell the future has always been very strong in
human societies. However, forecasting the future is not confined to
scientists. In a manner similar to scientific predictions, magicians
also use their hypotheses in an attempt to obtain some foresight about
the future. Modern man knows that predictions from scientific hypotheses have a much better chance of being correct than do the predictions from magical hypotheses, but remember that primitive man believes all his hypotheses are "scientific".

Magical foresight uses mostly "omens" or "oracles" to obtain information about the future. An "omen" is an object which humans
observe in order to give us clues about the future. Astrology is a
good example of an omen type of forecasting. By noting the positions
of all the heavenly bodies at a person's birth, astrologers purport to

predict the future for that person. Humans not only have tried to tell the future by looking at the stars, but also by observing other animals. The natives of Borneo will not build a new house until the flight of birds indicates approval. Many primitive tribes attempt to tell the future by inspecting the entrails of animals. Some tribes burn the bones of animals and then try to read the cracks produced by the heat, no doubt with about as much success as modern palm readers have in reading the lines in our hands.

It is also possible to tell the future by a mechanical oracle, that is manipulating an object as a type of "experiment". Modern examples are ouija boards and tarot cards. The Bantu peoples are able to obtain a great deal of information by casting several different animal bones on the ground. By observing the relationships between the various bones an expert can forecast a great many future events. At various times and places people have fed poison to chickens, speared bulls and watched which side the animal fell on, thrown grain into the air, cast bones, sticks and palm nuts, dropped hot wax into water, tried to read tea leaves and so on. The number of ways in which man has tried to predict the future is limited only by his imagination.

By definition all magical hypotheses are currently believed to be false. The question which many readers have doubtlessly been asking themselves is, if early humans were our intellectual equals, how could they have gone on believing such "ridiculous" hypotheses for thousands of years? A prediction made from an invalid hypothesis should prove incorrect a large proportion of the time. How could intelligent humans have gone on accepting these invalid hypotheses for such a long time? This is just the question the first anthropologists to study this area pondered (Tylor 1891 and Frazer 1890). In short, they proposed that individuals were unable to perceive the falsehood for a variety of reasons. For example, (1) many sick people treated with magic do get well; after a rain dance has been performed, it will sooner or later rain and so on. (2) In some cases magic actually "works", although for different reasons than are generally believed. Men have actually died when they became aware that a magician was attacking them with a spell. (3) Many magical spells are quite complex and detailed. It is believed that the charm must be repeated in exactly the correct order for the magic to work. When failure occurs, the magician could always claim that the spell had not been performed perfectly. (4) Undoubtedly the magician, either consciously or otherwise, in order to protect his own interest became quite adept at finding excuses for his magical failures. For example, he could claim that someone else's magic had counteracted his magic and so on.

However, Levy-Bruhl (1923) and Evans-Pritchard (1933 & 1934) critized this "intellectualist" interpretation by pointing out that people in primitive societies accept magical beliefs for the very same reasons we accept our scientific beliefs--because we were taught since child- hood that these ideas are valid. Most people accept, without much thought, the beliefs of the society in which they were reared. When you or I go to the doctor, we probably think something like "he is the expert and therefore he must know what he is talking about". If the medicine he gives us does not help, we probably think "perhaps he made a mistake in diagnosis or perhaps the medicine should have been stronger" or such. If after the second or third visit, we are still not any better, we might even think "perhaps I should try another doctor". A person in a primitive tribe is in exactly the same situation. He goes to the "witchdoctor" or local healer thinking "he is the expert and he must know what he is talking about". If the native doctor's medicine does not work he may question the diagnosis or the strength of the medicine or perhaps eventually even the ability of that particular na- tive doctor, but he is most unlikely to challenge the entire medical profession and their knowledge by proposing that this medicine is com- pletely useless! People in primitive societies often criticize and are sceptical of the details of the system, but seldom question its funda- mental assumptions. We are exactly the same.

Even among the experts (whether doctors or native doctors), there is a strong tendency to accept uncritically the major assumptions which they were taught as students. Hundreds of doctors in Europe accepted, apparently without a thought, the atmospheric-cosmic-tell- uric explanation of childbed fever. For every Semmelweis, there are hundreds of experts who calmly accept what they are taught without a murmur.

Only a very few people in any society are genuine sceptics. Most men are believers--they accept as valid whatever they are taught as children. When all the learned and intelligent members of a society agree about something, a young person is likely to accept that they must know more about it than he does, and having formed an opinion, most people are loath to change their minds. Or as Mark Twain put it, most people only think they think.

Certainly Levy-Bruhl and Evans-Pritchard's point is correct in so far as any given *individual* is concerned, as few people reason out their beliefs for themselves, but rather simply accept what everyone else accepts. But we still have the question: how could these early *societies* have gone on accepting these invalid hypotheses for such a long time? In our society, there are a very few people who are able to question and challenge the basic assumptions of their profession. Because of these rare geniuses, we are able to discard hypotheses which are faulty. We abandon inadequate hypotheses because a few brave individuals have the courage to point out to the rest of us how silly they really are! If our knowledge is able to progress in this way, why didn't it work the same way in early human societies?

There is another problem however which hindered early man in rejecting magical hypotheses, and this is also an important difference between magic and science. First it is important to realize that, as Evans-Pritchard (1965) pointed out, magical hypotheses seldom blatantly contradict experience. For example, few if any magical beliefs are as crude as, say "If you put your hand in the fire, it will not burn". Rather, magical hypotheses are almost always worded more like "If you put your hand in the fire, it will not burn *if you have sufficient faith.*" Magical hypotheses almost always have an "out" or a built in escape clause to explain failures. They are worded in such a way that no matter what happens, experience always seems to confirm the hypotheses.

Evans-Pritchard in his famous *Witchcraft, Oracles and Magic among the Azande* (1937) described an actual example of this among the Azande tribe of Africa. The Azande attempt to foretell the future by feeding a certain poison to fowls. The poison kills some chickens within a few minutes, but other chickens are apparently unaffected by it. The Azande ask the question beforehand, being careful to state which of the two possible outcomes means "yes" and which means "no". Usually they ask questions concerning the health and safety of their family. One of the most common questions asked is "Will I die this year?" They then feed the poison and wait for a result. The chicken will either succumb or survive the poison and thereby supply the answer. If the oracle predicts, for example, that the person will die this year, the individual is not really too upset as he has a variety of magical preventive measures he can employ to change this prediction.

An interesting feature of this oracle is that on important issues, a single test is insufficient. To be valid the oracle must give the same prediction on *two* tests, on one test the fowl must die, but on the other it must live. Obviously, the poison often kills or spares both fowls, so it would seem to us that this would be a clear sign that there is something wrong with the whole proceeding. However, Azande belief is not so simple. They have several secondary elaborations or ad hoc hypotheses which are readily pulled out to explain any failures. The apparent failure could have been due to (1) the wrong variety of poison has been gathered (2) breach of a taboo, (3) witchcraft, (4) anger of the owners of the forest where the creeper grows, (5) age of the poison, (6) anger of the ghosts, (7) sorcery, (8) use.

In other words, magical beliefs are seldom testable or falsifiable. Since magical hypotheses, like scientific hypotheses, propose relationships between separated events, it might at first seem that magical explanations would usually be testable, but this is not the case. Evans-Pritchard concludes that the Azande mystical notions "are eminently coherent, being inter-related by a network of logical ties, and are so ordered that they never too crudely contradict sensory experience but, instead, experience seems to justify them" (1937, p. 150).

In short, if we had been born and reared among the Azande, this poison oracle would probably seem just as reasonable and logical to us as it does to them. Magical beliefs only seem "ridiculous" to us because they are so far removed from what we believe. Our beliefs probably seem just as ridiculous to them as theirs do to us. The oracle *seems to work*, there are no obvious discrepancies or faults in the system which their magical theory cannot explain. And even if there were a rare truly sceptical Azande about, it would be very difficult for him to find a flaw in the theory.

In a way, Semmelweis was lucky, as the atmospheric-cosmic-telluric influence theory was, at least on one point, falsifiable. Remember that, by chance, his hospital was divided into two clinics and that these two clinics differed drastically in their rate of childbed fever. Semmelweis realized that this difference could not be accounted for by the theory all his colleagues accepted. How could this vague epidemic influence plague the first clinic but not the second, Semmelweis asked. He saw that the theory could not account for the evidence. A sceptical Azande might have an even harder time overthrowing the poison oracle theory than Semmelweis had in overthrowing the atmospheric-cosmic-telluric theory.

This gives us our first clue why early humans took so long making scientific discoveries. Magical theories are seldom falsifiable, so it was much harder for their rare geniuses to see and then prove a fault in the accepted theory. Modern science makes a conscious effort first to consider, and then to accept only those hypotheses and theories which are testable and falsifiable. It should be possible, at least in principle, to prove our scientific theories wrong, but this was not the case with early humans' magical theories. They couldn't abandon their magical hypotheses because they couldn't prove them wrong.

We are still left with the question: why was this the case? Why are magical theories seldom if ever falsifiable? The answer to this question will have to wait for a moment, however. First it is necessary to discuss the second type of belief prevalent in primitive societies.

RELIGION

Although magicians may sometimes believe in gods or spirits, they do not seek favours from these beings. Magic always assumes that man himself is capable of controlling his world. With the correct spell, the magician maintains that humans can directly manipulate nature to their liking. Religion, on the other hand, denies that humans have this power but always assumes the existence of some type of supernatural spirits that are usually more powerful than man and to which man may humbly direct his requests. Magic then, assumes that man can directly control nature, whereas religion denies this but asserts that there are other more powerful (but unseen) beings in the universe which can sometimes be persuaded to help us out. Thus both magic and religion assume that man has some sort of control over his destiny, magic proposing a direct control, whereas religion accepts only an indirect control through the help of unseen spirits.

Frazer proposed that in early human tribes magic most likely preceded religion. Magic is, after all, much simpler than religion, it does not necessarily assume the existence of unseen beings. Magic assumes that man is in charge of his own destiny and this would have had a tremendous psychological appeal to early man. Religion on the other hand, requires man to demote himself to a secondary power, to admit that he does not have the ability to control nature.

Today's anthropologists reject Frazer's claim that magic necessarily preceded religion. We have no solid evidence to support Frazer's proposal. Anthropologists have no way of knowing what the first humans believed. Today's primitive peoples invariably have a mixture of magic and religion. However, there are some primitive tribes that rely mostly on magic. It is possible that because of its simplicity and appeal, early man *may* have relied primarily on magic in his attempts to control nature, but there is no real evidence at this time to support such a theory.

There are probably few words as difficult to define as "religion". Religion is defined here as a system of belief that proposes the existence of beings or spirits whose existence is impossible to test with empirical evidence. Like magic, it is almost impossible to test religious proposals. Admittedly this definition is relatively narrow, particularly as in recent times much broader definitions of religion have sometimes been used which could conceivably label almost any strongly held belief system as "religious". However, defining religion as a belief in spirits or gods is not only the accepted traditional definition, but also the standard definition used in anthropology since Tylor (1871). Therefore even though some people would prefer a broader definition, the reader is requested to bear in mind that I am using the word in its traditional and narrow sense.

Usually these proposed religious spirits or beings are believed to have the power to control nature in some way. Spirits or gods come in all shapes and sizes. It is virtually impossible to list or even classify all the different beings which humans have believed in. Some tribes believe in many gods, each of whom has a specific function. For example, the Ifugaos of Luzon believe in around 1,500 spirits. Other societies believe in fewer but more general gods. Even in modern societies which often claim to believe in only one god, there are usually a variety of lesser spirits such as devils, demons, angels, saints, elves, ghosts and such in which many people believe.

Although some of today's primitive tribes rely mostly on magic, all primitive societies believe in the existence of "souls", that is that some essential spiritual core of a person continues to exist even after the person has died. Belief in the existence of souls is of course a belief in afterlife. As discussed in the previous chapter, it seems likely that humans did not propose the existence of souls until after men had accepted the hypothesis that life must end. It was not necessary to believe in an "afterlife" until humans had finally accepted that death is inevitable.

It is at least possible then that the belief in souls was one of the first religious beliefs. It is possible that before man proposed the existence of souls, humans believed that life could be prolonged indefinitely by the proper magic. With the proper spell, life need never end. The acceptance of inevitable death must have been delayed many hundreds or even thousands of years by hopeful magical proposals that death could sometimes be avoided. The final painful acceptance that the magical assurances of never-ending life were false, led early men to propose the existence of unseen souls that continued to exist after death. Where magic had eventually failed to give man eternal life, religion succeeded.

It is true that there is only some "circumstantial" evidence to support this idea that the first humans believed they might live forever.

This evidence consists of (1) humans only began burying their dead about 70,000 years ago. (2) Some of today's primitive tribes believe that all human deaths are caused by either (a) accident, (b) suicide or (c) sorcery, the implication being that if you could somehow avoid these hazards, you might never die. (3) Young children are not aware that they are going to die. But, as with Frazer's theory that magic preceded religion, there is no way for today's anthropologists to know what early humans believed.

However, science only requires that, to be considered, theories should be testable *in principle*. For example, if someone proposed that there are violent "earthquakes" on the planet Pluto, this would not be a testable hypothesis today, but it is quite possible that this proposal will be testable in the future. As yet, there are parts of Einstein's theory of relativity which scientists have been unable to test mainly because of the very high velocities with which his theory deals.

There is at least one experiment which *could* be conducted to test both Frazer's theory and my suggestion that early humans believed they might live forever, although I would oppose conducting this experiment on moral grounds. The experiment would be similar to the one suggested in Chapter 6 (p. 150) to prove that sexual intercourse is instinctive in humans. Basically it would consist of placing several young human children in an isolated environment (an island?) without adults and leaving them for years, perhaps generations to see what type of beliefs they developed. There would be enormous practical difficulties with such an experiment such as making sure that the children would remember very little if any of our culture, but nevertheless making sure they would be able to survive without adult help, as well as making sure they did not suffer any emotional deprivation in the process. And I want to repeat that I would strongly oppose conducting such an experiment on moral grounds, my only reason for mentioning it here is as part of a logical argument that such an experiment could *in theory* be conducted, therefore both Frazer's and my proposals are testable *in principle*.

Also, it is possible that in the future scientists will develop other tests which could be used to examine these ideas. The first anthropologists could not have dreamt of the radioactive carbon 14 dating test, for example, with which we can now make a reliable estimate of the age of objects. Likewise, it is conceivable that other new as yet undreamt of techniques will some day be developed which would throw light on this subject of early human beliefs. Both Frazer's and my proposals are falsifiable *in principle*.

Religion also makes an attempt at foresight. Magic and science make predictions based on their hypotheses concerning events separated by time. Since religion does not propose true hypotheses, the nature of religion's attempt to foretell the future differs somewhat from magic and science. Religious predictions come from men known as "mediums" or "prophets". These men purport to communicate in some way with whichever spirits that particular society happens to believe in. It is claimed that the gods give information about the future to the mediums or prophets who then pass this on to the rest of the community. However, religious predictions of the future are notoriously inaccurate. For example, at least as of today, exactly 100% of the many thousands of religious predictions by various prophets that the earth will end on a particular day have been found to be invalid.

Although magic most closely resembles science in that both propose true hypotheses, in another way religion resembles science. Religion proposes the existence of unseen beings, likewise scientific theories sometimes propose the existence of objects which cannot be perceived even with the aid of instruments. As briefly mentioned in Chapter 5, science sometimes assumes the existence of "proposed entities". Scientific theorists find that in order to relate the relevant events in a meaningful way, it is sometimes necessary to propose the existence of an object or process that cannot be observed. For example, when first proposed, atoms, molecules, electrons, bacteria and viruses were only "proposed entities"; however, many of these objects have now been observed through the use of improved instruments.

If both science and religion propose the existence of objects that cannot be perceived, is it possible to distinguish between the scientific and the religious proposals? Usually science has proposed the existence of inanimate objects, forces or minute organisms, whereas religion almost always proposes either the existence of powerful spirits which take an active interest in humans or else the existence of human "souls". However, there is nothing in principle to stop science and it is possible that someday science will propose the existence of some powerful living beings which cannot be observed.

To be classified as a possible *scientific* proposed entity, however, it must be possible to test by some indirect means the nature and the existence of this object. Religious proposals are always described in such a way that it is impossible to support or not support the existence of these beings with empirical evidence. Religious people usually strongly resist any attempt to subject their spiritual beings to any sort of empirical test.

Scientific proposed entities should be falsifiable. Although I have listed several examples of scientific proposed entities which are still assumed by scientists to exist and thanks to improved instrumentation some have now actually been observed, it is also possible to draw up a list of historical scientific proposed entities which scientists no longer accept. Some examples are phlogiston, ether, and fomites. These entities were eventually rejected by scientists because improved empirical evidence no longer supported the assumption that they existed. On the other hand, religious proposals are never rejected for empirical reasons. Of course, societies sometimes change the gods they believe in, but this is always done for a variety of other reasons.

A historical example of an entity proposed by a scientist and how this proposal was supported by empirical evidence is shown in the work of Torricelli, who had been a secretary and assistant to Galileo. Before Torricelli, it was known that by use of a suction pump it was possible to draw water up the pump barrel to a maximum of about 10 metres. It was not understood, however, why this maximum height existed. Galileo had proposed an explanation, but this was found to be invalid.

In 1643 Torricelli proposed that this phenomenon could be understood if it was assumed that the earth was surrounded by a "sea of air" which exerted pressure upon the surface of the earth. The weight of the air pressed down on the surface of the water and forced the water up the pump barrel when the piston was raised (thus removing the air pressure inside the barrel). The maximum height of about 10 metres resulted because at that height the weight of the air equalled the weight of the water in the barrel.

Now at least in Torricelli's age, this "sea of air" was not observable. Our bodies are constructed so that we cannot feel the weight of the air. Torricelli's sea of air was only a "proposed entity". However, by a series of experiments, scientists were able to provide empirical evidence to support Torricelli's non-observable proposal.

By conducting an experiment using mercury in a glass tube instead of water, Torricelli showed that the mercury only rose to a maximum height of about 75 cm. By conducting this experiment, Torricelli invented what we today call a barometer. The results of this experiment supported Torricelli's theory because mercury is about 14 times heavier than water. Since the sea of air is pressing down the same amount on both the water and the mercury, Torricelli's theory would propose that the maximum mercury height would only be 1/14th that of water, that is $10/14$ = about 75 cm.

More experimental evidence to support this theory was found by Pascal. He assumed that if Torricelli's theory were true, then the weight of the air would gradually become less as the altitude was increased, that is the higher the altitude, the less the weight of the air still above you. By repeating the experiment at the top of a mountain, this proposal was confirmed, the height of the mercury was 75 mm less than that measured at the bottom of the mountain at the same time.

Thus, although Torricelli proposed the existence of something that could not be perceived in his day--a sea of air whose weight presses down on the earth--it was nevertheless possible to provide by a series of experiments some empirical evidence to support the existence of this proposed entity.

To be classified as a possible scientific proposed entity, the object must be defined so that it is possible to bring empirical evidence at least in principle to support or not support its existence. In short a scientific proposed entity must be falsifiable. Even though scientific proposed entities are not observable, it is always possible to support their existence through the use of what have been called, by different writers, "bridge principles" or "correspondence rules". These "bridge principles" connect the theoretically assumed entities with the observable events. With Torricelli's theory, one of the bridge principles was that the heavier the liquid used in the experiment, the shorter the maximum height of that liquid. The weight and the height of the liquid are observable events.

It is difficult to discuss the nature of these bridge principles in general. They vary according to the nature of each theoretical proposed entity. The important point is that the proposed entity must have a definite structure or organization which will allow scientists to conduct tests using a bridge principle in several varying conditions. These empirical results must then vary as predicted by the theory before the theory will be accepted by scientists. To be classified as a possible scientific proposed entity, the theory must have at least one bridge principle, thus making the theory falsifiable.

Thus, although both science and religion propose the existence of objects which cannot be observed, scientific proposals can always be distinguished from religious ones by bridge principles. Scientific proposed entities must have at least one bridge principle, but religious proposals never do.

Although religious entities are never testable, it is not possible to claim that, therefore, they necessarily do not exist. In Chapter 5 it was suggested that non-testable hypotheses cannot be assumed to be

invalid. Several non-testable hypotheses have eventually become testable and then scientific hypotheses. Likewise non-testable proposed
entities cannot automatically be classified as non-existent. It is only
possible to claim that these religious proposed entities cannot be classified as *scientific* proposals.

Thus from a scientific point of view, religious proposals can never
be accepted nor rejected. A hypothesis or theory cannot be accepted
as scientific unless it is falsifiable at least in principle. Therefore
religious beliefs cannot be accepted as scientific; scientists cannot
classify these proposals as non-existent either. Non-testable hypotheses and theories fall outside the boundaries of scientific endeavour;
their existence cannot be judged by scientific inquiry. The possibility
that they exist is always there, but empirical evidence can never be
used to support this possibility.

Religious explanations, like magical explanations, never blatantly
contradict experience. Other than the fact that you can never observe
these spirits, their proposed existence *seems to be* a reasonable
enough assertion to the average person. Their existence seems to be
supported by empirical evidence. If you pray for it to rain and sure
enough it rains, this "proves" the god(s) exist. If however, it fails to
rain (for a long time), there are various ad hoc hypotheses which explain why the god(s) did not deem it necessary to fulfil your request.
Perhaps someone had broken the commandments and the god(s) are
punishing us or such.

Religious beliefs, like magical beliefs, are seldom if ever testable
or falsifiable. No matter what happens, empirical evidence always
seems to support these proposals. So much for the argument that
science is more "empirical" than magic and religion. Rather, science
is distinguished from magic and religion precisely because it is *less*
empirical. It is sometimes possible to prove scientific hypotheses
wrong. Empirical evidence *always* supports magic and religion, but
this is not so with scientific ideas. We are able to reject scientific
hypotheses precisely because empirical evidence does *not* always support them. (It is still possible to argue, however, that science is more
empirical in the sense that it is possible to *reject* scientific proposals
on the basis of empirical considerations, but that this is seldom if ever
possible with magic and religion).

Recently in the field of anthropology and sociology there has been a
lively debate as to whether magical and religious proposals in primitive societies are always "rational" (see Wilson, 1970). Some anthropologists have argued that, when considering whether this or that prim-

itive culture's beliefs are "rational", we should define rationality as the natives themselves define it, rather than impose our definition on them. Others have argued that if we do this, obviously all societies must surely believe themselves to be "rational" in some sense of the word, and that therefore the term would become virtually meaningless. Surely in some sense, our scientific beliefs must be "more rational" than their magical and religious beliefs.

I have tried to avoid the use of the word "rational" in this book because it is such a vague and ambiguous term which can be given many different meanings. Nevertheless, the following points can be made:
(1) All belief systems in different societies are "rational" in the sense that if you were born and reared in that society they would seem to be confirmed by empirical experience. Any reasonable person reared in that society would probably accept them as valid. There is never a blatant flaw in logic or conflict with experience from the native's point of view.
(2) As Gellner (1962) has pointed out, even when anthropologists are able to detect a flaw in a native theory, this flaw may be performing some important function in their society for instance maintaining class divisions. This "flaw" may be necessary for their society to continue to run smoothly. All belief systems (including our own) probably have some flaws which an outsider could easily detect. An anthropologist visiting the United States in George Washington's day would have no doubt noticed the logical contradiction between the existence of the institution of slavery and the famous statement in the American Declaration of Independence that "all men are created equal".
(3) No belief system is "rational" in the sense that it is possible to justify or prove the validity of a hypothesis or theory. We know that even some of our "scientific" ideas will undoubtedly be rejected in the future because new evidence will contradict their proposals. It is always possible that some new evidence will crop up in the future to dethrone even the most prestigious scientific theories. All hypotheses must be accepted on a certain amount of "faith", as it is never possible to prove a given hypothesis true.
(4) Scientific theories may be "more rational" than magic and religious ideas in the sense that it should be possible to reject scientific theories on the basic of empirical evidence, whereas it is seldom if ever possible to reject magical and religious hypotheses because of empirical considerations. Thus, only science is able to progress by chucking away faulty theories.
(5) Scientific proposals are more rational than magical and religious

ones in the sense that modern societies which have accepted the idea
that hypotheses should be falsifiable are obviously much more able to
manipulate and control their environment. It does not take an anthropologist to see that modern societies are more efficient at controlling
the environment than are primitive societies. Whether this ability is
good or bad is another question, but our ability to fly across the oceans
or to the moon or dam mighty rivers or survive in the antarctic or
eradicate smallpox is undoubtedly evidence that the scientific outlook is
in some sense "better" than the magical or religious orientations.

It is now possible to return to the question of why did early human
societies take such a long time making scientific discoveries? It has
already been suggested that early societies relied primarily on magic
and religion and that these types of belief are seldom if ever falsifiable,
that is, it is impossible to prove them wrong. Early societies had
trouble rejecting their magical and religious beliefs because it was
virtually impossible to prove them wrong. The beliefs seemed to be
supported by empirical evidence, so that even a rare sceptical person
would have trouble finding a major flaw in the argument. No one would
bother proposing a new hypothesis or theory to explain a particular
phenomenon unless they had first decided that the presently accepted
explanation must be faulty.

The important question remaining is: why was this the case? Why
are magical and religious beliefs always non-testable? The simplest
possibility is that the first humans just didn't know any better. It is
possible that the scientific outlook is a "learning set" which has gradually been accepted by humans over the centuries. . It is possible that
we gradually learned that certain types of hypotheses turn out to be
more useful than other types of hypotheses. We gradually learned how
to learn separated relationships. There may be something to this idea,
but even if so, it doesn't really tell us very much. We are still left
with the question why did it take us so long to acquire this "learning
set"?

A more promising theory was suggested by Horton (1967). In essence he proposed that the scientific outlook is encouraged by an "open"
outlook in a society, that is where members of that society are forced
to recognize that other alternative belief systems exist. This open outlook is encouraged by such things as the advent of writing, which as
Goody & Watt (1963) pointed out, forced literate societies to recognize
that their own ancestors had different beliefs from themselves. Pre-
literate societies are unable to realize that their own beliefs are slowly
changing over time. Any awareness of the small alterations or innova-

tions continually made in a society's belief system are eliminated by
selective recall by adults when these beliefs are passed on by word of
mouth to the younger generation. Adults only pass on a single consistent set of beliefs, forgetting or eliminating any small changes made
during their lifetime. But with the advent of writing, the ideas of
former generations are "frozen" in writing and people are forced to
recognize that their ancestors accepted different ideas from those accepted at the present time. This awareness forced people in literate
societies to be more sceptical of their own beliefs.

It was only about 2,500 years ago (in sixth-century BC Greece) that
the first truly literate society existed. Although writing was first developed about 5,000 years ago, the first forms of writing were all
pictographic. Pictographic systems required such a long period of
training that only a small elite of trained people were able to read and
write in each society. The newer phonetic alphabet was much easier
to learn and allowed, for the first time, a majority of the citizens to
become literate.

The "open" outlook was encouraged not only by the advent of writing,
but also by other factors such as the increasing trade and exploration
between different societies. Meeting people from other societies who
have completely different belief systems would also force at least a
few people to examine their own beliefs far more critically. Trading
which occurred in small cosmopolitan communities where people of
different cultural origins lived next door to each other was far more
effective in fostering this awareness than was the earlier type of trading where the people met in a neutral area, exchanged goods and then
returned to their own society. The first of these trading communities
occurred, again, in certain parts of sixth-century BC Greece.

The "open" outlook forced people to critically examine their own
beliefs. Through either writing, trading or exploration people became
aware that there are other possible belief systems. Even though the
beliefs in your society have always seemed logical, reasonable and
have never been contradicted by experience, if you are suddenly made
aware of other, different explanations of the world this may well force
a critical reappraisal of your beliefs. Even though your own beliefs
may be non-falsifiable, if you learn of other non-falsifiable belief systems, this may, at least in a few individuals, result in a sceptical
reappraisal of our own ideas.

Undoubtedly there is a great deal of truth in Horton's proposal. Exposure to foreign belief systems undoubtedly stirs critical thinking.
However, his ideas can only be regarded as a partial explanation of

why early man made so few scientific discoveries for the following reasons: (1) his suggestion does not even attempt to explain why the hypotheses and theories of early societies were always non-falsifiable in the first place. He only explains how it was easier to break out of these systems once alternative beliefs became known. (2) Since the first truly literate and trading society did not occur until about 2,500 years ago, his proposal cannot account for the increase of scientific discoveries which occurred *before* this time. As discussed in the previous chapter, science existed long before the advent of writing. And at least in relation to what little had happened during our existence up to then (2-4 million years), there was obviously a considerable speedup in scientific discoveries beginning about 10,000 years ago. The commencement of agriculture, domestication of animals, probably the sex-pregnancy discovery, smelting of metal, the making of pottery as well as the advent of writing itself all occurred in the *relatively* brief period of about 5,000 to 7,500 years. Clearly our rate of scientific discoveries had already speeded up considerably *before* literacy. Horton's theory has something to say about the dramatic speed-up *after* 2,500 years ago, but what about the relative speed-up *before* this time?

Actually, Horton's idea only reflects the "ego-historical" bias with which we still view science. Because we do not have any detailed account of what happened before the advent of writing, we have assumed that very little, if anything, must have happened before this time. But on the contrary, if we could make a chart of the *rate* of major scientific discoveries from the first humans to now, it would show how nearsighted this view really is. Such a chart might look something like this.

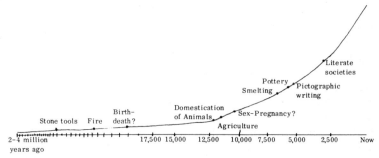

So what then can be the explanation? Why did we have such difficulties for such a long time? Why were the theories of early men always non-falsifiable? I shall argue that early humans unconsciously made their hypotheses non-falsifiable precisely so that they *could not*

be disproven, that early humans unconsciously did not want to know
that their ideas might be wrong.

First, consider the following points:

(1) Although there is a tremendous variability in the magical and religious beliefs found in primitive societies, there is one aspect of their
belief systems which is always the same. They never, or almost never,
admit that they are ignorant on a matter which they consider to be of
great importance. Of course, they sometimes admit their ignorance
on matters they consider unimportant, but if they are asked why did
they have a crop failure last year or what causes a serious disease
commonly found in their tribe, they will always give an answer of some
sort. They seldom if ever say "I don't know" to a question which is
important to their survival. Anthropologists have usually assumed that
they were reluctant to admit their ignorance on these matters because
to confess ignorance would be to admit to themselves that they were
powerless to control problems which affect their health, safety and
survival and to do this would arouse extreme anxiety. When confronted
with a problem that affects our very survival, it is very difficult if not
impossible for humans to admit to themselves that they are completely
ignorant and helpless. When faced with imminent danger, humans
would sooner believe in anything than in nothing.

Even when primitive tribes do sometimes admit their ignorance on
important matters, they often show clearly that this is most disturbing
and troubling to them. For example, with the recently discovered
stone age Tasaday tribe in the Philippines, after anthropologists had
repeatedly probed and questioned them on many subjects, one member
of the tribe said "We seem to get a wound inside us when people are
asking, asking, asking. They want us to make answers, but we only
know what we know" (Nance, 1975, p. 234). It was only after their
first contacts with modern civilization that these people became aware
of many areas of their ignorance. The helicopters, radios and numerous questions showed the Tasaday how limited their own knowledge
actually was and this was very distressing to them.

Modern scientists can admit to themselves that they are ignorant
about important questions because they live in a vastly safer world.
Modern scientists don't have to worry about where their next meal is
coming from or if there will be another plague next week. Admitting
our ignorance on important questions is a luxury which is afforded only
to humans who are relatively free from fears about their survival in
the immediate future.

Since primitive societies accept primarily magical and religious

explanations, and since they are unwilling to admit their ignorance on
matters of great concern, this means that magical and religious explanations are usually concerned with matters that are of vital importance
or worry to humans.

For example, Malinowski (1948) found that primitive people used
magic only when they had no scientific solution to the problem and the
matter was of particular concern or danger to the natives. When the
Trobriand Islanders went fishing in the calm lagoon where there was
very little danger to the fishermen, magic did not exist. But when the
same fishermen went fishing in the open sea where danger and uncertainty existed for them, there was an extensive magical ritual in an
attempt to secure safety and good results. Primitive tribes dispense
with magic when the phenomenon is either not of particular importance
to them, or else when they have sufficient scientific knowledge to deal
successfully with the problem.

Religion and magic are often strangely silent on subjects that are
not of particular concern to humans and our survival. Remember that
the Trobriand Islanders also had an elaborate religious theory which
proposed the existence of spiritual beings to explain how humans reproduced, but when Malinowski asked them how pigs reproduced, they
answered that "the female pig breeds by itself". By this it was meant
that unlike humans there were no spirits involved in pig reproduction;
the female pig reproduced by herself without the aid of spirits or males.
When Malinowski suggested that perhaps small pigs, like human babies,
were brought by their own spirits, the natives were unconvinced and
uninterested. Apparently they had no particular reason to believe in
the existence of pig spirits. Human reproduction was of great importance to the natives, but pig reproduction was not.

Likewise, virtually all religions propose the existence of human
"souls" or spiritual afterlife for humans. However, many religions do
not bother to propose the existence of an afterlife for other animals.
Those religions which believe in reincarnation sometimes involve the
idea of afterlife for other animals, but in these cases it could be argued
that this is only because it is believed possible for humans to be reborn
as animals. Religion proposes the existence of afterlife for humans
because mandatory death is a subject of great importance and anxiety
to humans, but whether other animals have any afterlife is not of particular interest to humans. Believing that other animals have an afterlife would not provide any particular emotional comfort for humans.

Although science also shows some bias towards phenomena that are
of importance to humans, science is by no means exclusively interested

in human related subjects. The breeding of the duckbilled platypus, the width and composition of the rings of Saturn and the migration of whales are just three typical examples of the many subjects in which science is interested. Unlike magical and religious proposals, scientific proposed entities are sometimes advanced to explain events that have very little interest to most of humanity. For example, the fact that water would rise only 10 metres in a pump barrel was not of great interest to most people at the time Torricelli proposed his theory.

Although the fact that water would only reach a maximum height in a pump barrel had been known for some time, religion had not concerned itself with this mystery because it was not a source of worry or importance to mankind. If this phenomenon had been considered important by humans, no doubt religion would have proposed the existence of a "water god" or some similar being that prevented the water from rising any higher. Religion always steps in to explain those phenomena that constitute a source of anxiety, fear or concern to human beings.

Not only do magic and religion in general deal only with matters that are of vital interest to humans, but also individual people tend to become more magical or religious when they are in stressful situations. The saying that "there are no atheists in the foxholes" is at least in some cases true. When faced with a dangerous situation in which a person has little if any control over his destiny, the desire to find help and comfort in magical rites or powerful supernatural beings is strong indeed. Magic and religion provide hope when all else has failed, or to quote Pope, "Hope springs eternal in the human breast."

It seems likely that societies, like individuals, make more use of magic and religion in stressful situations. Even a very primitive society such as the Tasaday, who live in a warm tropical forest with relatively few dangers confronting them, make much less use of magic and religion (and are more willing to admit their ignorance) than do, for instance, the Eskimoes, who, though possessing a somewhat "higher" culture, live in a much more dangerous environment. Because the Eskimoes live in a much more precarious environment, there are many more stressful problems which they need magic and religion to deal with.

Thus, it is possible that the very earliest humans actually made less use of magic and religion than do many of today's primitive tribes. Before learning how to control fire, for example, early humans would have been restricted to tropical areas, and therefore areas usually abundant in vegetation. Only after man had tamed fire could be venture into colder and more hostile regions of the earth. Magic and religion

only deal with, for lack of a better term, areas of "stressful ignorance".
If the first humans lived in a lush forest with an abundance of food, had
peaceful relations with neighbouring tribes, had few natural dangers
and still believed that they might live forever, they probably would
have had a relatively small need for magic and religion. Magic and
religion do not of course propose answers to all the mysteries of society, but they usually propose solutions for at least those obvious and
vital problems confronting the welfare and survival of that society.
(2) As already discussed, magical and religious proposals are virtually
always non-falsifiable. Their hypotheses and theories are always constructed in such a way that no matter what happens, experience always
seems to support the proposal. Through the use of secondary elaborations or ad hoc hypotheses, virtually all possibilities of disproving the
theory are eliminated.
(3) Even though magical and religious explanations are virtually always
non-falsifiable, if a sceptically orientated person is somehow able to
think up an experiment which *would* test these ideas, this proposed experiment is invariably rejected as useless or unnecessary. For example, I once overheard an interesting, and in retrospect somewhat
amusing, conversation between two young men. The first man made
some particularly blasphemous remark; to which the second man
warned him that it would be best not to make such statements for fear
of being struck down by lightning as punishment by the god in which
most people in that society believed. The first man (who evidently did
not believe in this spirit) only laughed. This discussion continued until finally the first man proposed a test--that he would be willing to
stand out in the middle of an open space during a rainstorm and challenge and dare this god (or any other) to strike him dead with lightning.
If lightning did strike him dead, then this would support the existence
of this particular being; on the other hand, if he returned alive, the
religious man would have had to admit that this did not support the existence of this spirit. Even though the religious man had first proposed
that this spirit had this power and he said he was still absolutely certain of the existence of his god, he would not agree to the test. The
religious man gave no reason for his refusal. Although in this instance,
it might be argued that the test was refused in order to spare the sceptic's life, no matter what type of test is proposed, religious people
always reject any type of experiment as unnecessary.
   The fact that magical and religious explanations are always non-
falsifiable and the fact that even when a genuine test is proposed it is
rejected by the adherents as unnecessary, strongly suggests the pos-

sibility that these explanations were deliberately (but unconsciously) made non-falsifiable precisely so that they *could not* be disproven. If you must reject a theory because it has been disproven, this also means that you must admit to yourself that you are in fact ignorant of how to control this phenomenon. Rejecting a theory is equivalent to admitting your ignorance. And admitting their ignorance on important matters is one thing primitive people find extremely difficult. I am suggesting that primitive tribes made their hypotheses and theories non-falsifiable precisely so they would not have to admit their ignorance. If a theory cannot be disproven, then the people who accept it never have to admit their ignorance. Thus, non-falsifiable theories prevent anxiety from arising.

My argument is as follows:

Unlike all other animals, early humans were acutely aware of the dangerous conditions in which they lived. Because of their unique ability to hypothesize, early humans were painfully aware of the dangers they were likely to face tomorrow or next week. For example, after experiencing insufficient food and predatory animals every day for the past week, early humans could hypothesize that these same problems were likely to occur again tomorrow and the next day and the next. Because humans are the only animals that are aware of what will be likely to occur tomorrow, only humans worry about tomorrow. Other animals may live in just as much danger, but they cannot hypothesize, they cannot predict that these same dangers will probably confront them over and over again in the future. Other animals show every sign of being frightened during a near miss with death, but because they cannot hypothesize, they must go to sleep that night with no worries about the next day. Early humans were much more aware of their vulnerable existence than were any other animals.

Early man was just as capable of hypothesizing as we are; however, because of this ability he was aware of the many dangers that were likely to confront him in the future long before he knew how to deal with them effectively. Undoubtedly the first hypotheses proposed were invalid or magical since there are almost an infinite number of hypotheses to explain any one event. The chances of proposing the correct hypothesis on the first few attempts are extremely small. It seems likely that, as Frazer has suggested, most hypotheses proposed by early man were unconsciously based on the principles of similarity and contagion, that is, that the consequence either resembled or had once been in contact with the event.

Early man had started in the only way he could to learn separated relationships. He began proposing hypotheses. Instead of proposing his hypotheses strictly at random however, he often tried to propose hypotheses that had some sort of identifiable connection with the consequence. However, as we now realize, the next stage in learning separated relationships is to test each hypothesis carefully, and reject those that are not supported by empirical results. A new hypothesis is then proposed, tested and if necessary rejected. The process continues until a hypothesis is suggested that is supported by all the empirical tests available to us at the present time.

It is at this point that the process of learning separated relationships broke down for early man. Rejecting a hypothesis is equivalent to admitting your ignorance. For early man this would have meant admitting his ignorance on virtually all areas of his life for many generations. As has already been suggested, this would have been more than mankind's emotional capacity could have managed. Admitting his ignorance would have aroused intense anxiety.

When confronted with a situation where his accepted theory had not adequately accounted for the evidence, early man had two choices: either (1) rejecting his theory and therefore admitting his ignorance or (2) adding a secondary elaboration or ad hoc hypothesis which modified and insulated the theory from having to be rejected. Through the use of these secondary elaborations early humans were able to prevent having to admit their ignorance and eventually to modify and protect their theories to the point where they were virtually impossible to disprove. Even if the hypothesis or theory had started off as a testable proposal, it was soon modified to the point where it became a *non*-testable proposal.

The problem, then, which prevented early man from making more scientific discoveries was not in the proposing of hypotheses, but rather in the rejecting of them. Because early humans could not admit to themselves that they were almost totally ignorant of the world, they continued to believe for thousands of years in magical and religious explanations. However, even though these explanations prevented anxiety from arising in early man, the snag was that they also impeded the proposing of new hypotheses and the acceptance of valid or scientific hypotheses even after they had been proposed. The magical and religious explanations may have prevented anxiety from arising, but ironically they also hindered the formation and acceptance of other hypotheses which would have allowed man some actual control over or even elimination of the dangers that were so upsetting to him.

SCIENCE

Although both magic and religion resemble science in some ways,
it is possible to distinguish them from science. Both magic and science propose hypotheses in an attempt to explain and control nature;
however, science can be distinguished from magic because only science
requires that its hypotheses should be testable and falsifiable. Likewise, science also sometimes resembles religion in that both propose
the existence of "proposed entities" which cannot be observed. Again,
however, science alone demands that its theories should be falsifiable
through the use of "bridge principles" before a proposed entity is assumed to exist. On the other hand, religious proposals never require
such testing before they are accepted. Thus, magic can be seen as the
non-falsifiable antecedent of scientific hypothesizing and religion can
be seen as the non-falsifiable antecedent of scientific theorizing of
"proposed entities".

Science is distinguished from both magic and religion, then, by its
requirement of empirical testing before either a hypothesis or a theoretical proposed entity is even temporarily accepted. When hypotheses
or theories are "tested" we are not of course testing the validity of
nature, we are testing the soundness of our ideas about nature. Therefore an experimental test of a hypothesis is only meaningful if it is
possible to show that the hypothesis *might be* wrong. Science is in-
debted to Karl Popper (1934) for pointing this out. If a hypothesis is
constructed in such a way so that no matter what happens the "test"
will support the theory, then there is really no point in even conducting
the experiment in the first place. A "test" must be just that, it must
be capable of discriminating between an idea that corresponds with the
nature of the universe and an idea that does not. Magical and religious
explanations have a great deal of empirical verification, the problem
rather is that they have *too* much. They are never wrong.

If a hypothesis or theory is not falsifiable, it will never be possible
to abandon the theory because of empirical considerations. Science is
able to progress because its theories should be capable of being proved
wrong. Science advances by continually chucking away some of its
ideas which have been falsified and by adding new hypotheses which are
falsifiable.

Because it is possible to falsify some ideas (but it is never possible
to verify or prove with certainty that any idea is true), we can sometimes know that a given hypothesis is false at least at the present time,
but we can never be sure whether other hypotheses are, in fact, true.
Our certain knowledge in fact only consists of knowing that certain

ideas are false.

Science is more effective than magic and religion in controlling the environment because its experimental method is actually a mini-attempt to control the environment. An experiment is actually a very carefully staged attempt to manipulate the environment. Hypotheses which are shown to be capable of controlling the environment are accepted and used on a much larger scale by many people, whereas a hypothesis which is not shown to have this capacity is rejected and forgotten. Science may not be any nearer the ultimate truth than magic and religion, but because it is highly selective about the nature of the hypotheses and theories it will even consider, and because it requires that before these ideas are even temporarily accepted, they must be found effective at controlling the environment, science is of course much more effective at doing just that.

I have tried not to give the impression that all scientific theories are invariably falsifiable. Scientists agree that scientific theories *should be* falsifiable, but in practice this ideal is sometimes not lived up to. Some "scientific" theories such as those of Freud and Marx, for example, are not falsifiable. Also, scientists of course continually alter and attempt to improve their theories in light of new empirical evidence. There is a very fine line between *improving* a theory by modifying it and *protecting* a theory by adding ad hoc hypotheses. It is sometimes very difficult to distinguish between these two activities. Any falsifiable theory could easily be made into a non-falsifiable theory simply by adding a few ad hoc hypotheses. Scientists have decided that their theories should be falsifiable, but unfortunately in practice this is sometimes not the case. Falsifiability is really only a convention which scientists have decided to accept, but as in many other areas of human endeavour, practice sometimes falls short of theory.

Although today's anthropologists do not accept Frazer's suggestion that magic preceded religion, they do accept the second part of his proposal that modern societies are slowly relying more and more on science and less on religion. The gradual change in emphasis can be documented in the steady conversion from religious to secular in European societies in the last few hundred years. Also, societies in Africa and Asia which are currently undergoing rapid modernization are also experiencing the same trend towards secularization.

Magic, religion and science all attempt to explain our world and to provide some method whereby humans have some control over our destiny. As a consequence, in spite of denials by various scientists and religious leaders, these three forms of belief are always in more or

less open conflict with each other. Since they all attempt to explain
nature, if one of the three gains in importance in a particular society,
then one or both of the others must decline. The gains of one must be
made up with the losses of the others. For example, if a particular
individual accepts the magical explanation for a certain event, it is
most unlikely he will also accept the religious or scientific explanation.

As more and more scientific hypotheses are accepted by a society
the earlier types of hypotheses are consequently discarded. For example, most Trobriand Islanders now accept that pregnancy is caused
by sexual intercourse, hence they will eventually completely discard
their belief in procreation by baloma spirits. Belief in pregnancy by
these spirits will cease when the people accept the more useful scientific hypothesis. Most primitive tribes (as well as early European
societies) believe that illness is often caused by an evil spirit taking
possession of the person's body. However, when modern medicine is
introduced, the belief that illness is caused by these spirts is slowly
abandoned.

European history also shows many other examples of conversion
from religious explanations to scientific hypotheses. Darwin's theory
of evolution is the most recent example. Before Darwin, western civilization universally believed that the human species was created by the
Judaeo-Christian god. Darwin's theory proposed that all species orig-
inated by evolving from earlier species. This implied that humans had
also evolved from other species. Darwin's proposal required no spirit
to create mankind. At first there was a strong religious opposition to
Darwin's theory, but today his theory is accepted universally by educated people and even most religious people now accept his ideas. Before Darwin, Christians had accepted the literal explanation given in
the Bible--that the Judaeo-Christian god had created man out of dust--
after Darwin, most Christians accept this as only a metaphorical explanation. It is clear that the literal religious explanation was slowly
abandoned when a more useful scientific theory was proposed.

The process by which a new theory overthrows the established theory
in the scientific community has been referred to as a "scientific revolution" by Kuhn. Magic or religion make no such drastic changes on
the basis of empirical evidence. However, when entire societies go
through a somewhat similar conversion, this might be referred to as
a "belief revolution", when most of the members of a society change
from believing in one of the three types of belief (either magical, religious, or scientific) to one of the other two types. "Belief revolutions"
which mankind must have passed through in pre-history were switching

from the magical or religious explanations about the origin of plants
and animals to the scientific seed-plant and sex-pregnancy hypotheses
and also, changing from the magical proposal that life could be prolonged indefinitely by the proper spell to the religious belief in afterlife.

There are many other examples such as eliminating magic, evil
spirits and "god's will" from the fields of medicine and mental illness.
Most events important to humans such as earthquakes, economic de-
pressions, eclipses, lightning, bad weather, famine and so on at one
time had magical or religious explanations. The fact that most people
in our society now accept the scientific explanation for all these events
means that there must have been a "belief revolution" at some time in
the past for each of these events in which the magical or religious beliefs were abandoned in favour of the scientific. Very little study has
been done on exactly how a society makes one of these belief revolutions. Certainly a more detailed examination of this phenomenon
should be made. An ideal area for study would be in a primitive tribe
which has recently come into contact with our western civilization.

Although there is often talk about the "resurgence" of religion (or
recently even of magic), careful analysis always shows that these gains
are always in the areas where science still has no clear answer (that
is, either accepts no hypothesis or else accepts only a limited or incomplete hypothesis). In these cases, it is not so much that magic or
religion has gained on science, as that they have gained in areas where
science still admits its ignorance. For example, science still does
not attempt to predict the specific future for individuals. It is in precisely this area where much of the current "resurgence" of magical
proposals such as fortune telling, astrology, palm reading, tarot cards
and so on have been most successful. On the other hand, there has
been no resurgence of magical or religious explanations in the areas
of birth control, curing yellow jaundice, prevention of plagues, or
fertility rites to ensure abundant crops and so on and so on. At one
time magic or religion had explanations for all these events, but there
has been no resurgence in these areas because science has proposed
suitable explanations for all these phenomena.

Magic and religion always propose explanations for those troubling
events for which mankind still does not have a scientific explanation.
In early societies where man's scientific knowledge was extremely
small, this meant that magic or religion explained almost everything
that was considered important. However, in modern societies where
science now can explain many events, the role of magic and religion
is correspondingly reduced. It is upon mankind's abhorrence of help-

lessness and ignorance of future dangers that magic and religion feed.
As the areas of our ignorance slowly recede, so do the areas which
magic and religion control.

This is not to suggest that some day science will have all the answers
and that magic and religion will then vanish forever from the face of
the earth. It seems likely, if not probable, that there are some questions to which science will never be able to propose an answer. Questions such as the origin of the universe, exactly how life commenced,
the specific future of individuals, the purpose (if any) of our existence,
and if there is any type of afterlife are still a mystery to science today
and most of these questions are still a source of great concern to most
people. It is just these questions to which magic and religion still propose answers. Since it is highly unlikely that science will ever be able
to explain all such questions, there will always then be a limited role
for magic and religion. These two types of belief will continue to
shrink, but they will probably never die completely. Unless man, after
thousands of years of concealing his ignorance from himself, is suddenly able to admit his ignorance on such important questions, then
magic and religion will continue to exist and to fill a need.

Because of all the objections raised to classifying beliefs as either
"magical", "religious" or "scientific", many anthropologists would prefer to drop this system of classification completely. Extensive modifications to Tylor's and Frazer's original definitions and ideas have
gone a long way towards answering these objections, but perhaps a new
classification would encompass recent thinking more clearly. Perhaps
instead we could classify beliefs according to the following criteria:
1) Hypotheses and theories can be divided into those which a particular
society presently accepts, and those which it does not. Naturally
enough each society (including our own) always assumes that the beliefs
they accept are valid or "scientific" but that proposals by other societies or even beliefs which were formerly accepted in their own society
are invalid or magical or superstitious or outdated. But as already
discussed no society can ever be sure its beliefs are indeed valid or
"scientific".
2) Hypotheses and theories can be classified according to whether or
not these proposals are testable and falsifiable at least in principle.
Most of today's western scientists would agree that scientific proposals
should be falsifiable. Most of our scientific proposals are falsifiable,
but most magical and religious proposals are not.
3) Beliefs can be classified according to whether or not they propose
the existence of unseen objects or beings. Hypotheses propose a direct

connection between observable events, but theories often propose the
existence of unseen objects or beings. Magic and non-theoretical science make use of hypotheses, but religion and theoretical science propose the existence of unseen entities or beings. It seems likely that as
Horton (1967) has suggested, people resort to theoretical proposals
only after they have found their hypotheses to be clearly inadequate.
If this is so, this would support Frazer's (modified) contention that the
first humans relied mainly on hypotheses (i. e. magic) rather than on
theoretical proposals (i. e. religion).

It seems unlikely however, whether the terms magic, religion and
science can ever be abandoned completely even in the field of anthropology, much less among the general population. Thus, I will continue
to use these terms, but only in the highly modified sense discussed.

Since all newly born members of all societies (both primitive and
modern) must pass through a ten to twenty year period of "education"
and socialization in which they are taught the magical, religious or
scientific explanations of the world which are currently accepted by
their society, most people have already accepted these explanations as
valid before they reach the age at which they are capable of thinking for
themselves. So it is only the rare genius who is able first to reject the
explanation he was taught as a child or student and then to propose another hypothesis which is closer to the truth. The history of science
shows that often it is the rejecting of the currently accepted hypothesis
which is the most difficult part. This process of rejecting the currently accepted hypothesis is undoubtedly even more difficult when this
hypothesis is non-falsifiable.

Even after the scientific giant has proposed his new explanation, his
society is often very reluctant to accept it. Often extreme social pressure is used against men who challenge the established authority. Both
Copernicus and Darwin delayed publishing their theories for many
years because of their fear of public (mostly religious) reaction. Established scientific authority is also often hostile to radically new scientific theories. Remember the hostility and opposition which both
Semmelweis and Jenner encountered.

Remember that Malinowski was not successful in convincing the
Trobriand Islanders that sex caused pregnancy. If Malinowski was not
successful, then the scientific genius who first proposed this hypothesis thousands of years ago may well have also met considerable public resistance to his new idea. Unlike today's scientist who is able to
record his ideas in writing, if the pre-historic scientist was not able
to convince his tribe during his lifetime of the validity of his theory,

then his ideas were most likely lost for ever. Historical science has
shown several examples where new discoveries have been "ahead of
their time" or out of tune with current scientific thinking and consequently ignored and forgotten for many years until someone else re-
proposed the idea. The two most famous examples are Aristarcus
(who proposed a sun-centred system in the third century B. C.) and
Mendel. This problem must have been much more acute for a pre-
historic scientist because magic and religion held a much stronger grip
on these early societies and because he could not record his ideas in
writing.

So the road that a scientific genius must follow in all ages is rough
indeed. First he must challenge and then abandon the teachings of his
elders, second he must propose a new hypothesis or theory which is
more effective in controlling the environment than the currently accepted explanation, and last he must somehow convince his society to
accept his new and often radical ideas. At least in some cases, the
second step is undoubtedly the easiest of the three.

Since most people in all societies accept some sort of explanation
(either magical, religious or scientific) of events that are important to
them, the man who proposes a bold new theory will almost always find
that it is in conflict with the already accepted ideas. Magic and religion are almost always hostile towards new theories that clash with
their beliefs; however, science also often opposes new hypotheses and
theories. The scientific community often shows considerable hostility
towards new theories, particularly if they involve a radical departure
from the current trends.

Scientists are, after all, people. Scientists often develop strong
emotional attachments to a particular theory. A scientist may have
spent all his working life accepting and attempting to find more evidence to support a particular theory. When a bold new theory is proposed which conflicts sharply with the explanation he has accepted all
his life, often the scientist shows that he has a strong emotional com-
mitment to the old theory.

As a consequence, individual scientists can be just as unwilling to
reject their theories as any magician or priest. However, although
individual scientists are often very dogmatic, the scientific community
as a whole has repeatedly shown itself capable of abandoning one theory
and adopting a radically new one. Kuhn has called this process a "scientific revolution" (see Chapter 5). Often it is only the young scientists
who are able to switch from the old to the new theory, and the process
of conversion is not complete until all the older scientists have died.

So even though some scientific theories are not, in fact, falsifiable and even though some individual scientists are probably incapable of rejecting their theories no matter what new unfavourable empirical evidence turns up, it is still possible to argue (contrary to Barnes 1974, p. 31-2) that science as a whole is falsifiable. The process of rejecting one theory and accepting another may take many years, but the fact that science has repeatedly undergone this process proves that its ideal is often lived up to.

The process of rejecting an explanation which has been accepted for many years (be it magical, religious or scientific) is difficult indeed. It is only because scientists have accepted that their theories should be testable and falsifiable that science is able to make the painful switch at all. But whereas science finds the conversion extremely difficult, magic and religion find it virtually impossible. The many different magical and religious beliefs, of course, change from time to time, but these changes can usually, if not always, be traced to the influence of beliefs from other societies, influence from particularly strong leaders, political domination and so on. Only science changes explanations on the basis of empirical evidence. As a consequence, only science is able to progress.

Although it has been stated that only science is willing to admit ignorance on important matters, even science sometimes finds this process difficult. Even when a particular scientific theory begins to show clearly that it cannot account for all the present evidence, scientists are often reluctant to discard it at least until a better theory is proposed. The phlogiston theory of combustion is a good historical example. Even scientists loathe to face the unknown without some sort of theory upon which they can build.

Although magical and religious proposals prevent emotional insecurity, they also impede the proposition and acceptance of scientific hypotheses and theories. However, it should be noted that each new scientific explanation which is accepted by a society probably makes it easier for more scientific explanations to be accepted in the future. The success of scientific method ensures that future scientific proposals will meet with less and less resistance. In some sense science feeds on itself. It seems clear that the first scientific discoveries (in pre-history) were extremely difficult. On the other hand, particularly in the past two or three hundred years, scientific discoveries have become more and more frequent and widespread. It is probable that this increase in the willingness of western civilization to accept scientific explanations--even when they conflict with magical or religious explanations--will continue to grow in the future.

There are several reasons for this progressive growth of science. One factor undoubtedly is Horton's "open outlook" which developed as societies became larger and more complex. Another factor may have been simply the increase in the number of people. Hunting and gathering societies for example are capable of supporting only a very small number of people in a given area compared to agricultural societies. As human population increased, this also increased the possibility of a rare genius being produced. Another factor may have been the complete isolation of any intellectual and sceptically-minded person in a small tribe. In larger societies, such individuals are often able to seek each other out (in universities for example) and find some associates among like-minded people.

However, the most important factor which allows science to grow at an ever-increasing rate is that many scientific discoveries reduce mankind's fear and ignorance of the future and thus decrease our emotional uncertainty about what may happen in the future. Whereas early man realized many of the dangers that could befall him in the future but did not know how to take any preventive measures, modern man is able to avert many possible future calamities by taking the proper action now. For example, by planting new high yield crops, and encouraging the use of birth control, we are attempting to avert predicted future mass starvation due to a vastly over-populated earth.

Early man made so few scientific discoveries because his overwhelming emotional insecurity prevented him from admitting his ignorance. However, since many scientific discoveries reduce our fear of future dangers, each discovery allows mankind to admit more easily our ignorance in other areas. Admitting ignorance about virtually all the known dangers was more than early man could manage, but admitting ignorance on only a few problems is much easier for modern man. A drowning man in rough seas who knows he cannot swim will cling ever so tightly to any scrap of wood he can lay his hands on, but an excellent swimmer in calmer waters would be more inclined to pass up an obviously inadequate piece of wood and search for a bigger and more useful piece.

Modern man knows he can control and prevent many of the dangers which we are aware may face us in the future and at least a few scientists are therefore able to admit their ignorance on important matters. Modern man is much more able to admit his ignorance about the unknown only because we live in a much safer world. Science has been growing progressively faster and faster because our fear of the future

has been shrinking progressively faster and faster. Early man faced a
"Catch 22" (or perhaps the term "Catch 1" would be more apt) concerning scientific discoveries. Before he could make a scientific discovery
to help him prevent a future danger, he first had to admit his ignorance,
but he could not admit his ignorance because he was so afraid of the
future. Our escape from this "Catch 1" was very slow at first, but has
speeded up dramatically in the past few centuries. It is certainly reasonable to assume that this process will continue in the future.

   In conclusion, it should be clearly pointed out that, in all ages, scientific discoveries--and in particular those of a radical and revolutionary nature--have been made by a mere handful of men and women. In
all societies, almost everyone accepts as valid the explanations which
they were taught as children or students. It is only a very rare and
independently-minded person who bothers or is able to question the
basic assumptions and explanations which everyone else has accepted
without a thought. It is the lone sceptic who dares to challenge single-
handedly the hitherto universally accepted explanations who makes
fundamental discoveries. So even though the benefits of scientific discoveries are often enjoyed by all of humanity, it is only a tiny group
of people who make these discoveries.

   Horton quotes the example of an Ijo tribesman who had been told by
a missionary to throw away his old gods. The Ijo man replied, "Does
your God really want us to climb to the top of a tall palm tree, then
take off our hands and let ourselves fall?" Rejecting the beliefs which
we were taught as children takes a great deal of courage. The reason
so few people are willing to undertake this activity on their own is undoubtedly because of this fear of the unknown. The distinguishing mark
of a genius is probably not his great intelligence, but rather his willingness or perhaps even pleasure at doing what other people regard with
horror, that is, his willingness to jump completely alone into the black
abyss of our ignorance. Heisenberg likened the role of the scientific
genius to that of Columbus, "who had the courage to leave behind him
all inhabited land in the almost insane hope of finding land again on the
other side of the sea". But the rewards found after such a voyage are
often beyond the wildest dreams of the solitary explorer. Describing
his feelings after his own considerable scientific discovery, Heisenberg
said, "At first, I was deeply alarmed. I had the feeling that, through
the surface of atomic phenomena, I was looking at a strangely beautiful
interior, and felt almost giddy at the thought that I now had to probe
this wealth of mathematical structures nature has so generously spread
before me. I was far too excited to sleep. . ." Perhaps our rate of

scientific discoveries is increasing because as our world is becoming
safer and more controllable, there are, relatively speaking, more and
more people who have the emotional capacity to "climb to the top of a
tall palm tree" and voluntarily make the terrifying jump.

After a new theory has been finally accepted by a society, the society
often heaps honours and glory upon the discoverer. However, societies
seldom if ever offer any help or encouragement to these people earlier,
when they are challenging the accepted doctrine or when they first proposed their new theory. Since these people are challenging assumptions
which everyone else has taken for granted, there is a definite tendency
for society to react with scorn, hostility and even ridicule. Although
some bold new theories have been accepted by society relatively quickly
(such as those of Newton and Einstein), other (such as those of Coper-
icus, Darwin, Semmelweis and Mendel) took many years before society
finally accepted them. Of these last four examples, only Darwin's
theory was widely accepted by society during his lifetime.

Few if any societies encourage their children to question or to challenge established "knowledge". The sceptic, the original thinker is
virtually universally regarded at least initially with suspicion. All
societies to a greater or lesser extent glorify obedience, respect for
authority and conformity. Even in modern societies this trend is still
evident. For example, when Einstein was a university student one of
his professors complained, "You're a clever fellow! But you have one
fault. You won't let anyone tell you a thing." Because Einstein alien-
ated all his teachers with his independent mind and his distrust of
authority, he was unable to get an academic job after graduation and
was forced to become a clerk in a patent office. As any of today's col-
lege or university students is aware, a student who persistently asks
challenging questions in class is much more likely to have his grade
lowered rather than raised.

Although societies praise conformity and acceptance of their fundamental assumptions, it is precisely on the rare iconoclast--who first
dares to think and later to boldly declare to the world that he is right
and everyone else is wrong--that revolutionary scientific discoveries
depend. Great scientific advances depend on the small handful of people
who have the courage first to reject society's explanations and then
when necessary to stand completely alone and defend their new ideas
even in the face of strong public hostility and rejection. It is just those
rare individuals who push, prod, nudge and coerce the rest of reluctant
humanity down the road of civilization.

Chapter 9

The Unique Ability

The question of how human intellect differs from the intelligence of all other animals has been pondered by men throughout history. Many philosophers and scientists have believed that our intelligence differs only in degree from that of other animals. Other scientists have argued that our intelligence differs in kind from that of all the other animals. This book has proposed that our intelligence is indeed different in kind. I have proposed that humans have not just one unique ability, but rather two. We are the only animals able to symbolize and also the only animals able to hypothesize. Given that humans are capable of these two unique abilities, what if anything do these abilities have in common and how are they related to each other?

In Chapter 2 it was proposed that our representational art and symbolic language are based on our ability to symbolize. Although other animals are able to learn some of our symbols, they are not able to create their own symbols. By repeatedly pairing one of our symbols with an event, other animals are able to learn the connection. Say "heel" often enough to a dog when he is standing at your feet, and he will soon be able to learn the connection.

So by repeatedly pairing a symbol with the event, other animals are able to learn some of our symbols, however they are not able to symbolize, that is to create their own symbols. Symbolizing was defined as a decision to form or assign a relationship between two dissimilar events that have not been paired together previously (p. 53). Learning someone else's symbols and creating your own symbols are two completely different processes. Learning a symbol consists of simply learning a connection between paired events; symbolizing is assigning a connection between non-paired events.

Although the concept of contiguity had not yet been discussed in Chapter 2, this principle was at least implied in the definition of symbolizing. The phrase "paired together" at least implied that they are placed together *at the same time*. Now that contiguity has been fully explained and discussed, we can streamline our definition of *symbolizing* as deciding to form a relationship between two dissimilar and previously non-contiguous events. Therefore learning a symbol consists of learning a connection between *contiguous events*, whereas symbolizing is a decision to assign a connection between dissimilar and previously *non-contiguous events*. Other animals are able to learn our symbols, but they are not able to symbolize.

In Chapter 5 it was also proposed that our magic, religion and science are based on our unique ability to propose hypotheses. A hypothesis was defined as proposing a connection or relationship between two or more events separated by more than 60 seconds (p. 111). Other animals are limited to learning contiguous relationships, but only humans are able to learn non-contiguous or separated relationships by proposing true hypotheses.

Given that humans have these two unique abilities--to symbolize and hypothesize--how are they related to each other? By comparing the definitions of these two abilities, we see that they are remarkably similar.

Symbolize--to decide to assign a relationship between two dissimilar
and previously non-contiguous events.

Hypothesize--to propose a relationship or connection between two
or more non-contiguous events.

It should now be evident that these two uniquely human abilities are actually two aspects of the same underlying ability. There is a very basic similarity between deciding that a particular word and a particular object are related to each other (i.e. symbolizing) and proposing that two events separated by time have a connection (i.e. hypothesizing). Symbolizing and hypothesizing are two sides of the same coin, they both involve proposing a relationship between non-contiguous events. Our two unique abilities turn out to be two different manifestations of the same underlying ability.

Other animals are of course able to communicate with each other and to learn. However, their communications and learning differ in kind from ours. Other animals are able to communicate by means of genetic signals, gestures and perhaps even by learning some of our symbols. However, our communication differs from theirs because we are the only animals able to symbolize--which enables us to create our own languages and to produce representational art. By deciding that the word "cat" shall represent a particular four legged-animal, our ancestors created a symbol. The word "cat" is not similar to or imitative of a real cat. Also, it is most unlikely that the word "cat" had previously been contiguously paired with this animal. To symbolize is to propose a relationship or connection between two dissimilar and non-contiguous events.

Also other animals are able to learn, but their learning is restricted to contiguous learning. Only if the pellet falls a few seconds after the rat presses the lever, is he able to learn the connection. However, humans are able to learn separated relationships by hypothesizing-- proposing a connection between two events separated by time.

Our ability to symbolize and hypothesize marks us apart from the rest of the animals. Our representational art and language rest on our ability to symbolize, and our magic, religion and science rest on our ability to hypothesize. But since these two abilities actually involve the same underlying process, perhaps a single term should be desig- nated to refer to this process. It is tempting to simply refer to this process as "symbo-hypothesizing" or "hypo-symbolizing". However, even a hard-nosed scientist might flinch at referring to man's unique ability with such an awkward, hyphenated term.

Science has often taken a word from the everyday language and, by giving it a restricted and precise meaning, has used the word as a technical scientific term. Some examples are "velocity", "metre", "work" and "resistance". This is what I have chosen to do in this case. I will use the word "imagination" to refer to mankind's unique mental ability of symbolizing and hypothesizing. The word "imagination" is commonly defined as the process of forming mental images or concepts of what is not actually present to the senses.

Of course, it has previously been suggested that man is the only animal capable of imagination. For example, Bronowski (1965) pro- posed that "Man has a richer life of experience than the other animals, because his mind alone works consistently with images, and thereby endows him with (literally) a life of imagination . . . man has experi- ences which do not happen--that is, which are not outward events" (p 17). But using the standard definition of this term, it was never possible to prove that other animals were not capable of imagination. It is not possible to show by experimentation that other animals do not form mental concepts or images.

For present purposes the term "imagination" will be defined as pro- posing a connection between two or more events that are separated by time, that is when all the events are *not* present to the senses at the same time. Since no other animals have been able to create their own symbols and since they are not capable of learning separated relation- ships, we are justified in assuming that only humans are capable of "imagination". These relationships can be diagrammed as follows:

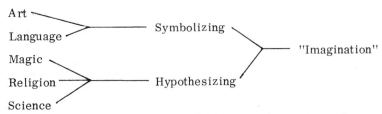

The term imagination is also useful because the traditional meaning
conveniently covers the process of advocating "proposed entities"
found in science and religion. Someone imagines an object, process
or being that he cannot actually observe. If experimental evidence is
found to support this proposal, then this is said to be a scientific proposed entity. If it is not possible to test for the existence of this
entity, then this is a religious proposed entity. Also artists often say
that before they begin they often imagine or form a "mental image" of
what they want their finished work to look like. The traditional meaning of imagination covers all these processes.

Since symbolizing and hypothesizing are actually two aspects of the
same underlying process, it can be safely assumed that our ancestors
acquired these two abilities at the same time. The unique capacity for
"imagination" was probably a result of one or more chance mutations
which resulted in the increased size and function of our brains. Since
this book proposes that human intelligence differs in kind from that of
all the other animals, it will be assumed that humans first existed
when our ancestors first became capable of "imagination".

In all likelihood, the first humans began symbolizing and hypothesizing almost immediately, and after they had reproduced and pro-
liferated, very quickly acquired symbolic language, some form of art
and proposed many hypotheses about their world. Very soon after
biological evolution had given us the capacity for "imagination", early
man must have developed a symbolic language and also proposed various hypotheses about his environment. However, it is at least possible that the earliest humans--like today's chimps--did not have an
adequate vocal apparatus to produce spoken language. They may even
have had to partially rely on some form of sign language. Further
mutations would have added this ability to produce a complex spoken
language.

Anthropologists have long known that among today's "primitive"
tribes, there is no such thing as an immature language or art form.
All cultures have a language that is sufficiently large and complex to
suit their purposes. Indeed, some languages from primitive tribes

are a great deal more complex than many "modern" languages. Also
no art form from these "primitive" tribes can be considered childlike
or immature. Although some tribes do not have two-dimensional representational art or all the forms of artistic expression we are accustomed to, all cultures have some type of aesthetic expression. Sometimes these art forms are less well-executed than ours, this is particularly true where their tools and techniques are inferior to ours,
however there is maturity and sophistication in the art of all cultures.
In those societies that have representative art, none of them can be
said to be childlike or immature.

Also, no primitive tribe can be said to be seriously lacking in hypotheses to explain events that are of great importance or worry to
them. All peoples are equally able to hypothesize. Although most of
the explanations proposed by the people in primitive tribes are what
we would call magical or religious, they are still explanations.

No peoples can be said to be primitive or immature in their ability
to symbolize or hypothesize. At least in relation to language, aesthetic
art forms, magic and religion there is no such thing as a *primitive*
tribe. These facts have led some anthropologists to suggest that
"primitive" cultures are in no way inferior to modern ones. They propose that although these tribes are, of course, vastly different from
ours, we should not brand them as primitive or inferior. Although different from ours, their cultures are of equal value and sophistication.

Although this position is perfectly tenable in relation to language,
aesthetic art, magic and religion, it is ridiculous to suggest that these
cultures are on a par with us in relation to science. This is the only
area where "primitive" tribes can be genuinely called primitive. Their
knowledge of separated relationships is obviously and demonstrably
inferior to ours. If early men and today's primitive men were and are
our equals in relation to language, aesthetic art, magic and religion,
why should they be inferior to us in relation to science? Why should
science--out of all these areas resulting from our unique ability of
"imagination"--be different?

The answer is that science alone requires an external check on our
"imagination". Language, aesthetic art, magic and religion allow man
to use his "imagination" unchecked or unverified by reality. When
early man decided that a certain animal would be referred to by a particular word, he had created a symbol and had used his "imagination".
But when symbolizing, there is no "right" or "wrong". Any symbol is
as good as any other. The English word "cat" is in no way a better or
a more correct symbol than is the Spanish word "gato". Both symbols

are equally useful. It doesn't really matter what we call this animal
as long as we call it something. There is not a single "true" symbol
for this animal by which we can compare and rate each culture's
attempt at naming this animal.

Because any symbol is as good as another, there is no such thing as
a primitive language. The same holds true for aesthetic art, magic
and religion. In all these areas, man could use his "imagination" as
he liked, any proposal he made was as good as any other. As a consequence, although cultures differ enormously in their language, aesthetic
art, magic and religion, they are all equally useful and accomplish
their function in each society with equal efficiency. It is also likely
that once early man had acquired the ability of "imagination", he quite
quickly developed comparably mature equivalents in all these areas.

But science is a different matter altogether. Only science requires
that man check his "imagination" against the real world. Only science
says that some proposals are better than other proposals. Only science
has a type of external comparision by which various proposals can be
rated and found to be either right, wrong, superior or inferior. Only
science says that our "imagination" must be tempered and restrained
by external reality.

Even though early and primitive tribes were and are able to hypothesize as well as we can, they did not quickly develop an advanced science because of the additional scientific requirement of empirical
testing. Although man's ability of "imagination" is sufficient to quickly
develop language, aesthetic art, magic and religion, it is not sufficient
to develop science. In addition to the requirement of "imagination",
science also has another requirement of empirical testing. The scientist must not only be able to propose hypotheses, he must also be
able to reject them when empirical evidence does not support their
validity. As with language, aesthetic art, magic and religion, science
is based on the ability of "imagination", however, unlike all these other
activities, science also requires a strong dose of scepticism, of rejecting those proposals that do not measure up to external reality.
Science requires that if our proposals do not conform to how nature
behaves, then we must abandon our proposals and admit our impotence
and ignorance. Only science requires that we sometimes admit our
powerlessness and ignorance. As already discussed in Chapter 8, this
requirement was very difficult for early man to accept primarily for
emotional reasons.

It is interesting to note here that although all societies have mature
forms of art, not all primitive societies have two-dimensional repre-

sentational art. Like science, representational art attempts to describe reality. In both science and representational art humans are
not allowed to use their "imagination" unrestricted. On the other hand
with purely aesthetic art, as with magic and religion, man can allow
his "imagination" to flow unimpaired by the real world. Hence representational art is "scientific" in the sense that it attempts to describe
the external world. The oldest aesthetic art so far discovered is about
32,000 years old. When early humans first began producing representational art--now estimated to be approximately 20,000 years ago--
this could be called a "scientific" discovery in the sense that man
realized that in order to produce this type of art, his "imagination"
must sometimes be narrowed, tempered and even rejected if it did
not successfully reflect nature.

Because today's science has become increasingly specialized and
technical, and today's art is often not strictly representational, the
similarity between science and representational art is now often difficult to detect. However, in an earlier and less specialized age the
similarity was probably much more self-evident. At least one of
western civilization's greatest artists was also one of our greatest
early scientists. Leonardo da Vinci probably never made a clear distinction between representational art and science.

There is strong evidence that writing first evolved from two-dimensional representational art. The first forms of writing were "pictographic" (see p. 58). Anthropologists generally assume that the existence of writing in a culture marks the beginnings of a civilization.
Thus, not only scientific discoveries, but also representational art,
writing and even civilization itself depended upon egocentric man
finally recognizing that the world exists independently from how our
"imagination" would like it to exist.

Some people may suggest that our advanced knowledge of separated
relationships is of questionable value, that perhaps this superiority
creates more problems than it solves. This is of course a debatable
point. I have a certain amount of sympathy with both sides in this
dispute. Certainly modern technology has created many problems for
us which were unknown to our ancestors. However, it is also undeni-
able that it has allowed us to live vastly safer, more comfortable and
longer lives.

There is no doubt that scientific advances create new problems as
a result of solving old problems. Even the seemingly unquestionably
valuable work of Semmelweis, Pasteur and Lister had its negative side
effects. Through the work of these men and their modern day descend-

ants, humans were able to prevent countless deaths due to disease and infection. Few people would question the value of these discoveries. However, such discoveries as these allowed many more people to live longer, and eventually this has led to a tremendous increase in the population of humans. By lowering the death rate, but not the birth rate, an increase of population was inevitable. We are now faced with worldwide over-population, and partly as a result, many thousands of people die each year of starvation. Before these discoveries were made, thousands of humans died of disease and infection, now thousands die of starvation. Has science really helped?

One solution to our problem might be to chuck away all of our modern vaccinations, antibiotics and so on--after a few good plagues and epidemics, our problem of over-population would vanish. But another solution would be to try to lower the birth rate--to bring the birth rate down to match the low death rate. However, effective birth control requires more science (both medical and social science), more knowledge of how to control our environment.

Admittedly science creates new problems, in fact, many scientific advances have resulted from attempts to improve weapons of war and have thus led to more efficient killing. However, all scientific advances, since they allow man to control his environment to a better degree, also have positive uses. And the new problems created by scientific advances are also potentially solvable. For example, concerning our modern problem of over-population, science has now made reliable birth control possible. But here again new problems are created; all the present methods of birth control are either less than 100% effective or else have undersirable side effects. But the world is not perfect, never has been and never will be. There will always be problems. Science will never be able to create a perfect world, but it certainly seems possible that it is capable of creating a somewhat *better* world by allowing humans to more fully control their environment. It is even possible that some day social science will discover the cause(s) and therefore the conditions necessary to prevent warfare among humans.

When people today decry all the problems created by our advanced technology, they often forget (or have never been fully aware of) what the world was like before these advances. They denounce over-population and pollution, but they still want the benefits of modern medicine, a warm home in the winter and modern transportation. They eagerly accept the benefits of scientific discoveries, but they denounce science for creating new problems. They often say that "science has gone too far" or that "it has got out of hand", but if given the choice, would any of these people choose to live 100 years ago, 1,000 years ago, or 10,000 years ago? I think not. Few if any people have become so disgusted with modern civilization that they have shed all their civilization and walked completely naked into the jungles of Brazil or New Guinea to "return to nature".

Humans often remember the past through a rosy haze. The good times are remembered, but the problems forgotten. I suspect that humans have often yearned to return to an earlier time when the problems were supposedly less. I would not be at all surprised if, after early humans had discovered how to use fire, a few individuals reminisced about the "good old days" before fire. No smoke in your eyes (the first form of pollution), no wood to gather, no accidental burns, no ashes in your food, they must have lamented as they sat warm and comfortable around this horrid discovery. They would, of course, have forgotten about the cold nights, the raw food and the long inky darkness after the sun set.

Not only are symbolizing and hypothesizing two aspects of the same unique human ability, but also we must use both our symbolic language and our hypotheses together in order to create a widespread change of behaviour among humans. In order to reap the full survival benefits of our unique ability to see far into the future, we must use not only our ability to hypothesize, but also our ability to symbolize. To be effective, man must use both of these abilities in conjunction with each other.

In Chapter 4, an account was given of Ignas Semmelweis's discovery of the cause and the prevention of childbed fever. After a great deal of hypothesizing and experimentation, Semmelweis proposed that childbed fever was not caused by the "atmospheric-cosmic-telluric" influence which was the explanation accepted by most of his colleagues. Rather, Semmelweis hypothesized that puerperal fever was caused by "putrid" matter". He ordered his students to wash their hands with chlorinated lime before making examinations. Results in his hospital supported his hypothesis.

But even after Semmelweis had proposed this hypothesis, his job was far from done. Even though he had found how to prevent childbed fever, few doctors outside his hospital knew of his discovery. Semmelweis had discovered the cause of childbed fever, but without the use of symbolic language to communicate this knowledge to others, his discovery had only very limited effects. Making the discovery that there was a connection between "putrid matter" and childbed fever was not enough; in order for this knowledge to be of widespread survival value to humanity, this hypothesis also had to be communicated or transferred to other humans.

Scientific discoveries are usually made by one person or a small handful of people. But in order for a scientific discovery to increase mankind's chances of survival, this knowledge must somehow be acquired by many, if not most, of the people working in this area. This very important function of transferring the knowledge is performed by our symbolic language. A scientific discovery by itself is of little importance; without the equally important function of spreading or communicating this learning to others, it would be virtually useless. Symbolizing and hypothesizing are not only two sides of the same ability, but also they must work hand in hand if science is to have any survival value to humanity.

It was in just this area of communicating his ideas that Semmelweis's tragedy and failure lay. He loathed writing and had a naive belief that the "truth will emerge". Because of his failure to effectively communicate his discovery, his ideas remained largely unknown outside his native country of Hungary.

Semmelweis had at first attempted to spread his ideas mainly by word of mouth. This of course proved to be an almost complete failure, but it is interesting to note that this method of communication would have been the only method available to pre-historic scientists. Both speech and writing are now part of our symbolic language. But writing is only a relatively recent acquisition. For most of our existence, any new knowledge could have been spread only by word of mouth. This is a relatively slow and inefficient method of spreading knowledge over large areas. One of the major reasons why pre-historic scientific discoveries--such as the smelting of metals--took hundreds of years to spread across the old world was that knowledge could only be passed on by word of mouth, usually from one village to the next and so on.

Written language, particularly after the invention of the printing press, is a much more effective means of relaying knowledge quickly and accurately. Not only can an article or book easily be shipped halfway around the world and millions of copies distributed simultaneously to anyone interested, but also the information remains intact and accurate, there is no problem of inaccuracies and omissions appearing as when information is passed by word of mouth.

Also written language allows communication not only to people now living, but also accurate communication to future generations. A quote by Clarence Day (1920) describes succinctly how the written word is able to communicate the knowledge, thoughts and feelings of one age of men to following ages.

The world of books is the most remarkable creation of man. Nothing else that he builds ever lasts. Monuments fall; nations perish; civilizations grow old and die out; and, after an era of darkness, new races build others. But in the world of books are volumes that have seen this happen again and again, and yet live on, still young, still as fresh as the day that they were written, still telling men's hearts of the hearts of men centuries dead.

In recent centuries written language has also acquired more and more of a function of storing present knowledge. As the volume of our knowledge steadily increased, it soon became impossible for individuals to learn or "store" all of this knowledge in their heads. Today it is impossible for an individual to know more than a tiny fraction of the total of man's knowledge. Written language not only stores knowledge for future generations, but also for the present generation. There is probably a certain amount of knowledge stored in libraries which no person actually "knows". The author of the article or book may have died or forgotten part of his writing, and no one else has read it recently. In such a situation, the information is stored and exists "independently" of humans. Through the use of writing, it is possible for the total of human knowledge to be greater than the sum of the information known by all the individuals. A certain amount of our knowledge exists only on paper.

Through the use of spoken and written language, the learning of one person can be communicated to all people. The ability to hypothesize would not be much use to us without the ability to symbolize. We must use both of these abilities together in order to enjoy the benefits of scientific discoveries.

Other species are able to transfer a certain amount of knowledge from one animal to another by imitation, gestures and genetic signals. Imitation and gestures are limited only to the higher mammals however (see p. 37). For example, one chimp can learn how to perform a task simply by watching another chimp performing this task. Non-human animals are able to share their learning only by means of imitation, gestures and genetic signals. These are very simple methods of communication and are quite restrictive in the type and complexity of information that can be communicated (for an example see p. 47-48). Other animals are limited to these primitive forms of communication because they are not capable of symbolizing.

Non-human animals are restricted to learning contiguous relationships because they are not able to hypothesize, also they are restricted to communicating even this contiguous learning to each other by only imitation, gestures and genetic signals because they cannot symbolize. They are restricted in both learning and communication because they are not capable of "imagination".

In closing, the foregoing discussions in this and previous chapters should not be taken to imply that other species of animals are completely incapable of changing their behaviour towards events that will occur far into the future. This is certainly not the case. As discussed in Chapter 6, other animals sometimes display behaviour which has survival value only in relation to the distant future. Some examples are migration, nest-building and the hoarding of food. However, experiments have shown that these behaviours are instinctive and not the result of learning. Young animals which have been isolated from adult members of their own species are still able to perform these activities. These instinctive behaviours evidently originate from and are controlled by the animal's genetic inheritance acquired from his parents. Not only does the young animal acquire his physical and anatomical characteristics from the genes and chromosomes of his parents, but also his instinctive behaviours.

In 1859 Charles Darwin proposed that animals change their physical characteristics and new species originate through biological evolution. Through mutation and natural selection, animals change their physical structure so that it improves their chances of survival. However, this change via biological evolution usually requires many generations to become widespread. Likewise, animals should also be able to change their instinctive behaviours through the very same process of biological evolution. If by genetic mutation, an animal acquires a new instinctive behaviour which improves his chances of survival, then this new behaviour will have a good chance of being passed on to the next generation. For example, if one member of a migratory species, because of a mutation, alters his migration in such a way so as to improve his chances of survival, then this change will probably be passed on to the next generation. On the other hand, a new instinctive behaviour that decreased chances of survival would have much less chance of being passed on to the succeeding generation.

Thus through biological evolution, all animal species are able to change their instinctive behaviours. Other species are able to change their behaviour towards events that will occur to them in the distant future, however non-human animals are limited to the process of biological evolution in carrying out these changes. Only humans have an additional and non-biological method of changing this type of behaviour.

There is an intriguing similarity between these two methods of changing behaviour. The following diagram shows clearly these similarities.

Type of behaviour	Animals using	Name of process	Source of variation	Method of spreading change	Result
instinctive behaviour	all animals	biological evolution	mutation	sexual reproduction	change of behaviour
separated or scientific learning	only humans	"cultural" change	hypothesizing	symbolic language	change of behaviour

With biological evolution, changes are introduced through a genetic
mutation in an individual animal. If this change has survival value,
the change slowly spreads over several generations to the entire population. With "cultural" evolution, an individual human proposes a
hypothesis, this hypothesis is spread by the use of symbolic language,
and if accepted as having survival value, results in widespread change
of behaviour. These two processes are actually quite similar. In both
processes the new variation originates in only one individual and the
change is then spread to others. Both processes result in a change of
behaviour.

Expressed in terms of our separated learning, a newly-born animal
with a mutant instinctive behaviour is a "hypothesis". And from the
time of birth until the animal reaches reproductive age, the animal's
life is an "experiment". If this change proves to have survival value,
the "experiment is "successful" and the "knowledge" is passed on to
others.

A scientific hypothesis or theory on the other hand, when expressed
in terms of biological evolution, is a "mutation". If this "mutation"
has "survival value" it will be able to "reproduce" and will eventually
establish itself as a "species" (a school of thought which is seldom able
to fully communicate with other similiar schools). Often there are
several such "species" competing for the same scientific area. If one
of these "species" is clearly "better adapted" to explaining this area,
it will eventually drive all the other "species" into "extinction". This
dominant "species" would then become, in Kuhn's scientific terminology, a paradigm. This "species" would continue to dominate the area
until a new "mutation" with a better "survival value" was produced.

In both processes, the sources of variation--mutations and hypotheses--are most likely to produce changes that do not have survival value. Most mutants turn out to be disasters and most hypotheses turn out to be invalid. However, occasionally a variation has survival value and is then passed on to other members of the species.

One very important difference between these two processes, however, is the speed at which widespread changes of behaviour can occur. Biological evolution requires several generations, the time in years varies from species to species depending upon the time it takes the young animals to mature and reach reproductive age. In some species, new generations are produced every few weeks, in others many years separate generations. Through separated or scientific learning, humans can change their behaviour in a much shorter time. Today, thanks to modern communications, it is conceivable that humans all over the world could, in an emergency, change their behaviour towards a particular event within a few days, certainly within a few weeks after a hypothesis had been proposed and tested.

Biological evolution is able to introduce changes in behaviour only between generations. Biological evolution does not allow individual animals to change their behaviour during their lifetime. Core instinctive behaviours are fixed or set at the time of conception and cannot be changed or modified during the animal's lifetime. Individual animals cannot change their core instinctive behaviours, but it is possible that each newly-born animal's instinctive behaviour will differ from that of his parents. In contrast to instinctive behaviours, the process of learning allows animals to change their behaviour within their lifetime. Only through learning can *individual* animals change their behaviour.

All individual animals, through contiguous learning, are able to quickly change their behaviour towards events which will occur in the next few seconds. All species, through biological evolution, are also able to slowly change their behaviour towards events which will occur in the distant future. However, only humans are able to quickly change our individual behaviour towards events which will occur hours, weeks and years in the future. Needless to say, this ability gives us a tremendous survival advantage over all the other species.

PART THREE

Chapter 10

Norms for Survival

Is the pious or holy beloved by the gods because it is holy,
or holy because it is beloved by the gods?

Socrates

An account of the life and work of Ignaz Semmelweis was given in
Chapter 4. Although Semmelweis had discovered the cause of puerperal fever in 1847, his ideas remained largely ignored or unknown outside his native Hungary until after his death. In his book written some
twelve years after his original discovery, Semmelweis wrote:

If this treatise had no other object than to establish our
Doctrine on an unshakable foundation, and to make perfectly
clear the sad error of the epidemic theory of puerperal fever,
if this only was our object, we might suitably bring our treatise to a close here.

But that alone cannot be the object of this treatise, for my
Doctrine is not firmly established in order that the book ex-
pounding it may moulder in the dust of a library: my Doctrine
has a mission, and that is to bring blessings into practical
social life. My Doctrine is produced in order that it may be
disseminated by teachers of midwifery, until all who practise
medicine, down to the last village doctor and the last village
midwife, may act according to its principles; my Doctrine is
produced in order to banish the terror from the lying-in hospitals, to preserve the wife to the husband, the mother to the
child.

This quote can easily serve as an introduction to the theme of the
rest of this book. This theme is that many scientific discoveries do
not merely "moulder in the dust of a library", but that they often "bring
blessings into practical social life". Scientific discoveries are often
not only knowledge for knowledge's sake, rather they are used to bring
about drastic improvements in our health, safety and happiness.

The problem to be considered in this chapter is that often spreading

the knowledge of scientific discoveries through the use of symbolic
language is not sufficient to bring about the desired ends. Even if the
new knowledge has been effectively communicated and accepted by the
people concerned, this is sometimes not enough to bring about widespread changes of behaviour. For various reasons, people will sometimes not consistently behave in accordance with what they understand
and believe is in their own best interest and the interest of society. In
these situations, the society must sometimes encourage or even require that the individuals behave in a way which is deemed to have long
term survival advantages for that society. For example, in 1855 the
Hungarian government ordered all hospital authorities to introduce
Semmelweis's preventive procedures. Note that the government did
not simply inform the hospital authorities of his discovery, it *legally
required* them to follow his procedures. Simply communicating the
information to them was not considered sufficient, society required
that they behave in a way which was believed to be in their best interest.
    Society not only uses laws to coerce its members to behave in the
desired manner however, it also uses a variety of other types of rules
such as morals, customs, taboos, standards and manners. Psychologists and anthropologists have often used the collective term "social
norms" to refer to these various rules of behaviour. A general introduction to the study of social norms can be found in Sherif (1936). The
nature of these "social norms" will now be discussed.

SOCIAL NORMS
    All societies, primitive and modern, have rules to govern and to
attempt to control the behaviour of their members. Anthropologists
have not found any society which is completely "free", that is which
does not have rules dictating how its members may or may not act. It
seems most likely that early human tribes also had various rules of
conduct for their members.
    Societies use an assortment of different types of rules or norms.
These different types of norms vary in their tolerance of misconduct
and also in the severity and type of punishment which is bestowed upon
the violators. It was Sumner (1907) who first made a distinction between the various types of norms used by society. The following is a
slight modification of Sumner's three basic types of norms.
    The simplest type of rules are the folkways, customs or manners
which are usually relatively tolerant of misconduct, the only punishment being a certain amount of social disapproval by most members
toward the violator. Second are the morals, mores and taboos in which

misconduct is less tolerated and punishment is not only in the form of
social disapproval, but also by supernatural or magical means. If such
a rule is broken, it is believed that the spirits or forces will send bad
luck or disease or destruction upon at least the violator, and perhaps
the entire society. Supernatural punishment also often includes a variety of undesirable consequences in the offender's proposed "afterlife".
A taboo is a rule in which the punishment is believed to be "automatic",
that is it is a magical hypothesis concerning a necessary consequence
of a particular activity. For example, the Trobrianders believe that
stealing fruit from outlying and unprotected fruit trees inevitably results in the onset of hideous diseases. With a taboo the punishment is
"automatic", whereas with a religious moral, the punishment is "op-
tional", its administration depending upon the judgement of the relevant spirits.

The last type of rules and the ones least tolerant of misbehaviour
are the "laws". Laws are the rules which society requires its members (at least in theory) to obey. It is believed that each violation
should be punished. Punishment consists of fines, imprisonment, phy-
sicial pain, mutilation or death at the hands of a socially authorized
agent. The agent is recognized as administering the punishment on
behalf of the community as a whole and not merely on his own. Consequently, revenge against the agent by the violator or his family and
friends is not allowed. A most interesting study of law in primitive
tribes can be found in Hoebel (1954).

There is of course, a certain amount of overlap of these types of
norms. Often, for example, a rule which is considered very important
by a society will not only be a "moral" but also a "law". Punishment
for misbehaviour comes not only from the spirits but also in the form
of force from the society. Most primitive societies rely more heavily
on morals or taboos for social control than on laws. As societies become more and more secular, the effectiveness of magic and religious
sanctions diminishes, and the role of legal controls increases. Also,
there is always a certain difference between norms in theory and norms
in practice. All societies have their pretend rules which they state and
believe to be immoral or illegal, but which in actuality are tolerated,
especially if the offenders are not too blatant in their breaking of these
rules.

All societies use these three types of norms. However, the rules of
one society may differ tremendously from those accepted by another
society. Behaviour which is freely allowed in one society, may be severely punished in another. For example, in many primitive societies

(such as the Trobrianders) pre-marital sex is freely allowed, whereas among other societies it is allowed only under certain restricted conditions. Last, most modern societies until quite recently have attempted to prohibit pre-marital intercourse completely, using moral sanctions. There is an unusually wide variation among the rules which different societies use to restrict the behaviour of their members.

If different societies vary so much concerning what they allow and what they prohibit, is there any common denominator which underlies this multitude of often contradictory regulations? Is there some explanation as to why norms differ so? A closely related question is: do these different norms have a common purpose? Why do they exist at all? If norms differ so, are they really necessary? Would it not be possible to have a society without any rules or norms? Also, what effect do scientific discoveries have on a society's norms? And last, but certainly not least, why are humans the only animals with laws, morals and other types of norms? The rest of this chapter consists of an attempt to answer these and other questions.

The question to be considered in the next section is: if norms in different societies differ so, do they have a common purpose?

THE FUNCTIONAL NATURE OF NORMS

Different cultures vary enormously, not only in their norms, but also in virtually all other aspects of life such as methods of food production, tools, kinship groups, belief systems, education and so on. Despite this wide variability among different cultures, anthropologists such as Malinowski (1944) have suggested that all these divergent practices have one thing in common: that is they are all *functional*, they all attempt to satisfy an organic or survival need of mankind. All cultures must cope with the same basic problems of survival. Men in all cultures must eat, sleep, procreate, eliminate waste and so on. Hence all cultures must provide methods and procedures for dealing with these survival needs. Malinowski listed seven basic biological needs of humans: metabolic, reproductive, the needs for bodily comfort, safety, movement, growth and health. So although cultures differ tremendously from one another, all these divergent practices are functional in that they are attempting to solve these survival needs of individuals and the society as a whole.

For example, in the area of social norms, one of the few rules which all societies share is the prohibition of murder. All societies prohibit one person from killing another person. (All societies also make certain exceptions to this rule such as in warfare, executions by law officials or in some subsistence societies the killing of unproductive members of society). This prohibition against murder is obviously a functional norm. Any society which allowed its members to kill each other freely would soon be in disarray.

However, this is one of the very few norms which all societies share, most norms vary widely from one society to another. Critics of the functionalist school have asked why, if the aspects of different societies are all functional, do they differ so? If all norms are functional as Malinowski and others claim, why are they so varied and even contradictory? For example, if all sex norms are functional, why do they differ so? As White (1949) has pointed out, constants cannot explain variables.

At least some of the variability of norms would be explained by the functionalists as due to differences in environments. For example, a tribe living in an arid or desert environment might well have a rule against chopping down palm trees, however another tribe living in an area of abundant water, food and trees would most likely not have such a rule. The norms in these societies would be contradictory, but because of different environments, both their norms would still be functional. However, this environmental explanation of variability only explains a small part of the widespread diversity of norms and other aspects of different cultures. As yet, the functionalists have not been able to answer this criticism adequately.

Because of this and other such criticisms, many anthropologists today consider functional explanations trite, circular and even trivial. No matter what unusual actions, institutions or beliefs were found by anthropologists in remote societies, the functionalists simply assumed that they *must be* functional and set about thinking up explanations of why they were functional. Everything was "explained" by its function, but any other activity you might care to name could also be "explained" by the exact same procedure. Probably the most critical view of Malinswski's brand of functionalism was put forth by Leach (in Firth 1957, p. 120) who proposed that the functionalist theories of Malinowski "are not merely dated, they are dead". However, this author happens to believe that with certain crucial modifications to Malinowski's theory, functionalism can again become a respected anthropological outlook. Or, with apologies to Mark Twain, "reports of the death of functionalism have been greatly exaggerated!" Some such modifications to functionalist theory are proposed in this chapter.

In addition to these doubts about functionalist theory, many anthropologists have doubts as to whether *all* norms are actually functional.

They agree that *most* norms appear to be functional, but point to the
many examples of norms in different societies which certainly seem to
be non-functional or even dysfunctional. Some examples are the Hindus' prohibition on killing cows, and the Jews and Moslems' refusal to
eat pork.

This has led to an interesting controversy in which anthropologists
favourable to the functional school have attempted to explain how these
apparently non-functional norms are actually functional. For example,
Harris (1974) has put forward some very interesting arguments suggesting why several such apparently non- or dysfunctional norms in
different societies are actually functional for the survival of these
societies.

Harris's argument concerning the Hindu prohibition on killing cows
is briefly as follows. India is now and has been for some time a poor,
agricultural society. Unlike farmers in richer countries, Indian
farmers do not have the capital to buy tractors or artificial fertilizers.
However, cows and oxen in India provide low cost substitutes for tractors and chemical fertilizers. Oxen provide the only available means
of ploughing the fields and India's cattle provide about 700 million tons
of recoverable manure each year. Although there is an abundance of
cows, there are actually too few oxen in India.

The important point is that female cows are "factories" for producing
male oxen. Without oxen to plough the fields, the people would undoubtedly starve, and only cows can produce oxen. Add to this the crucial
factor of India's erratic weather. During droughts and famines caused
by the recurrent failure of the monsoon rains, the farmers would be
strongly tempted to kill their cattle for food. But even if the farmers
survived the drought, they would then be unable to plough their fields
or even produce more oxen when the rains did come.

Slaughter of cattle by farmers for food during drought would result
in a much greater danger to the Indian society after the rains came. A
few more farmers might survive because of the meat of their cattle,
but a great many more people would starve the following season because there would be no crops. Not killing cows even if you are starving is a functional norm because female cows produce not only milk
and manure, but also the mainstay of India's agriculture: oxen.

Such arguments proposing a hidden function or purpose for seemingly non-functional norms often seem reasonable and may well have
some validity, but there are many such apparently non-functional norms,
and anthropologists have not even attempted to explain how many of
these are actually functional. Before considering the question of wheth-

er *all* norms are actually functional, it will first be necessary to consider several other elements in this problem.

First of all, it is necessary to stress two points which are often neglected in this debate. First, to my knowledge no one has proposed that all norms are functional for the reasons *stated by the people* in that society. For example, no anthropologist has suggested that the Hindu prohibition of killing cows is functional for the Indian society because, as the Hindus claim, "cows are sacred". Rather these anthropologists have tried to show that such seemingly irrational beliefs are actually useful to the society for reasons of which most of the inhabitants themselves are not fully aware. The reasons people give for their norms cannot always be taken at face value. It seems that in many societies the vast majority of people are not aware of the true functional reasons for their norms. An explanation of this phenomenon will be given later in this chapter.

Secondly, it is also important to realize that anthropologists have not claimed that all norms must be functional for objective environmental or ecological reasons. Some norms may be functional only in the sense that they provide emotional comfort for the members of that society or a focus for group solidarity and so on. In other words, not all norms are necessarily functional in the sense that they help provide external environmental control and benefits such as food and shelter. Some norms may be functional only in the internal or subjective sense. Some norms may only serve a "subjective" or emotional purpose.

For example, in Chapter 8 it was mentioned that members of the Zande tribe in Africa put the juice from one of their medicine plants on their young crops in the belief that this makes them grow better. We know that this juice is ineffective as a fertilizer and therefore label this practice "magic". To my knowledge, no anthropologist has suggested that this practice is actually functional in an "objective" sense, in that it enables the tribe to control the environment in some way. However, it is commonly assumed that such practices are functional in the "subjective" sense, in that they reduce worry and anxiety in the people concerning the outcome of their crops. The Zande themselves believe that this practice is functional in an "objective" sense (i.e. it makes their crops grow better), we know this is not the case, but we can also see that this belief is nevertheless functional for "subjective" reasons (it reduces their anxiety).

Functional anthropologists have claimed that all norms are functional, but this is not to say that all norms are necessarily "objectively" functional. This point has not always been fully appreciated. It is important to bear in mind this distinction between "objective" and "sub-jective" function, as it will be discussed further in this chapter.

The next section proposes an answer to the problem of why norms differ vastly from one society to another.

NORMS AND HYPOTHESES

As discussed in Chapters 5 and 8, all cultures have explanations or hypotheses concerning most events that are considered important for their survival. In primitive tribes most of these explanations are of a magical or religious nature, however. At the beginning of this chapter it was also mentioned that all cultures also have a variety of rules or social norms which attempt to regulate the behaviour of their members. However, the norms used in one society may differ tremendously from those adopted in another society.

I now wish to propose that there is an important link between the hypotheses which a given society has accepted and the social norms which that same society uses. All explanations (either of a magical, religious or scientific nature) propose that a relationship or connection exists between events separated by time. In other words, most hypoth-eses propose that certain events occurring now may have an effect on what will happen in the future. Most hypotheses make predictions con-cerning what may happen in the future.

All human societies attempt to satisfy the survival needs of their members. However, societies are not only concerned with the imme-diate needs of their members. Because all cultures accept hypotheses (and hence predictions) concerning important events, all societies as-sume that they have a certain amount of knowledge and therefore pos-sible control concerning the conditions which will exist in the distant future.

A society is not only concerned with the immediate survival of its members and itself, but also with the long-term survival. All hypoth-eses imply that the future can be controlled or at least predicted to some extent. Since the hypotheses which a given society accepts imply at least partial control or knowledge of future conditions, human soci-eties can attempt to ensure not only the immediate survival of their members, but also, through hypotheses, the long-term survival of that society as well.

One of the ways societies attempt to ensure survival in the distant future is through the use of social norms. A society's social norms always attempt to control the behaviour of the members in such a way as to ensure favourable conditions for survival in the future. A society

always discourages activities by its members which are believed (on
the basis of hypotheses) will lead to conditions detrimental to survival
in the future, and societies always encourage behaviours which are
believed to result in conditions favourable to survival. For example,
among the Aines a pregnant woman is prohibited from spinning or
twisting ropes for two months before her delivery. This may seem to
us a rather silly rule and it is difficult to see how this norm could have
any survival value. However, the Aines believe this prohibition to be
useful because they accept the validity of the magical hypothesis upon
which it is based. The hypothesis is that spinning or twisting rope by
a pregnant woman causes her unborn child's intestines to become entangled like a thread. This prohibition is perfectly reasonable if you
accept the validity of the underlying hypothesis. A social norm is
always based upon a hypothesis accepted as valid by that society.

Thus, a great deal of the variation between norms in different societies can be attributed to the wide range of magical and religious hypotheses accepted in these societies. In addition to differing environments, societies differ in their norms because they differ widely in
the hypotheses they accept as valid.

Concerning the question of whether all norms are functional, it follows then from my proposal that at least all norms are *believed* to be
objectively functional by the leaders of that society. Since all societies
accept certain hypotheses as valid, the social norms which result from
these hypotheses would also be assumed to be valid or functional. They
assume that their norms result in "objective" benefits because they
also accept the validity of the hypotheses upon which they are based.
It is doubtful if very many people in a given society are able to see that
one of their own norms is actually only functional in a "subjective"
sense. The ability of a magical or religious hypothesis and its resultant norm to provide subjective or emotional comfort depends upon the
person assuming that it is "objectively" valid. Once a person seriously
doubts the objective validity of a given hypothesis and norm, then the
"subjective" benefits also vanish. For example, if a member of the
Zande tribe ever seriously doubted the objective validity of putting medicine juice on his plants (i.e. it makes them grow better), then he
would also lose the subjective benefits of this hypothesis (reducing
worry and anxiety).

The complicating factor here is that of course, although all societies
assume their hypotheses to be valid, not all hypotheses *are* valid. Magical hypotheses are invalid by definition and it is never possible to test
religious explanations to ascertain their possible validity. Thus it

would at first seem that although all norms are believed to be objectively functional by the members of that society, whether a given norm
was actually objectively functional would depend upon the nature of the
hypothesis on which it was based. If a social norm was based upon an
invalid magical or religious hypothesis, then it would seem to follow
that the resultant norm would be either objectively non- or dysfunctional. On the other hand, if a norm was based on a valid or scientific
hypothesis, then one would think it would result in a truly objectively
functional norm.

But the problem is not so simple. If we accept Harris's and others
arguments that some norms can be objectively functional for reasons
of which the vast majority of the people in that society are unaware
(and this certainly seems to be true in some cases), then it follows that
it is sometimes possible for a norm based on an invalid magical or
religious hypothesis to be believed objectively functional by society,
but nevertheless to be objectively functional for *other* reasons unknown
or only dimly perceived by most of the people. For example, the Hindus believe that their rule prohibiting the killing of cows is functional
because "cows are sacred" and therefore that killing these animals
would displease the gods; but Harris has argued that this norm, although based on a religious hypothesis, is nonetheless objectively functional for other reasons. Also, remember the Trobrianders do not
steal fruit from outlying and unprotected fruit trees because they believe that this results in the onset of a hideous disease in the thief. We
can easily see that this norm is based on a magical hypothesis and is
not functional in preventing diseases. However, we can also see that
this norm is objectively functional in that it reduces the amount of
theft in their society.

Is it possible then that norms based on a magical or religious hypothesis could nevertheless be objectively functional for reasons unknown
to most or even all of the people? If this is indeed the case, how did
this strange situation arise? An answer to this question will be posed
in a later section.

The problem of why norms are necessary in the first place has not
yet been discussed, however. If all the people were informed that a
particular activity would result in undesirable conditions for their survival in the future, would not this be sufficient inducement for them to
change their activities? Would not people voluntarily modify their behaviour so as to maximize their own survival in the future? Why must
society always coerce or force people to act in ways which they already
know to be in their own interest?

LEGITIMIZATION OF NORMS

All social norms are based on a hypothesis that certain effects will result from a particular activity. Assuming that a particular hypothesis has been widely communicated within a given society, it would at first seem that the individual people would eagerly adjust their behaviour in light of this hypothesis in an attempt to create favourable conditions for themselves in the future. Unfortunately, it is not always the case. For several reasons, people will sometimes not consistently change their behaviour in the direction indicated by the prevailing hypothesis.

The first and simplest reason for this is that some people may not accept the validity of the hypothesis in question. For example, in modern societies today there is a widely publicized hypothesis that smoking cigarettes causes lung cancer from ten to twenty years later. (There is also a certain amount of empirical evidence supporting this hypothesis). Even though most, if not all, cigarette smokers know of this proposal, some of these people do not accept the validity of this hypothesis. Some people just don't believe it. For justifiable or personal reasons, it seems likely that all societies have had at least a few sceptics who have doubted the validity of widely accepted explanations. Obviously, such a sceptic would be unlikely to change his behaviour in the direction indicated by the hypothesis.

The number of such true sceptics in any given society is probably very small, however. The real problems arise among the vast majority of people who unthinkingly accept a society's prevailing hypotheses, but nevertheless do not consistently change their behaviour. There are at least two interrelated reasons explaining the behaviour of these people. The first of these is that all hypotheses by definition propose that certain consequences will result from a particular activity *in the distant future*. Cigarette smoking results in lung cancer in ten to twenty *years*. The proposed consequences of all hypotheses are in the future, from as little as a few hours to as much as many years. Even though people know and accept the validity of hypotheses, the consequences are often too far removed in time from the present to cause them much concern. Even though cigarette smokers may know and accept the cancer hypothesis, the long time gap before the consequences appear makes it difficult for them to stop smoking. This is especially the case when the immediate (contiguous) effects of the behaviour are pleasurable. Cigarette smokers apparently obtain a great deal of immediate pleasure from inhaling smoke into their lungs, but the proposed undesirable consequences are many years removed. The contiguous desirable consequence (pleasure) is felt much more strongly than the distant undesirable consequence (cancer). This explains why many cigarette smokers have a very hard time giving up smoking. The immediate and strongly felt pleasure of the activity is set against the long-term and remote possible disease. The individual accepts that the same activity has both desirable and undesirable consequences, but he has a hard time changing his behaviour because the immediate effects are always felt much more strongly than the separated effects.

This same conflict is also evident when the separated consequence is desirable, but the contiguous consequence is undesirable. For example, most students accept the hypothesis that the long-term consequences of their education will be desirable, consisting of such things as a useful skill or the ability to obtain a well-paid job and so on, but the immediate effects of studying are often boring, tiring and just plain hard work. The student often has a hard time making himself study because the immediate consequences are felt much more strongly than the long-term consequences. Even though people may accept the validity of a given hypothesis, they sometimes find it difficult to change their behaviour because the distant consequences are too remote and abstract to strongly motivate them. This problem is greatly increased when the immediate effects of this same activity create a conflict situation between desirable and undesirable consequences.

The second factor which compounds this difficulty people sometimes have in changing their behaviour even when they accept a given hypothesis, is that although the immediate consequences of a certain activity may only concern the individual, the long-term consequence will concern not only that individual, but also many or even all of the other people in that society as well.

For example, mainly because Britain is an island, she is now free from the scourge of rabies in the wild animal population which is rampant on the main continent of Europe. On the continent rabid foxes, dogs and other animals continuously bite and infect humans, but because of the English Channel, the people in Britain are spared this danger. Although animals do not swim across the English Channel, many (mainly pets) are brought over in boats and planes by their owners. Britain's long standing law requiring pets brought in from abroad to spend six months in quarantine has recently been strengthened and the punishments for people who attempt to evade this required isolation for their pets have been increased. This is obviously a functional law in that it attempts to prevent the spread of rabies to Britain.

However, many people still attempt to evade this regulation. The

problem is that although the long-term desirable consequences (prevention of rabies) concerns 55 million humans (and an unknown number
of animals), the immediate undesirable consequences affect only the
individual pet owner. The owner is of course very reluctant to have
his beloved pet put in a kennel for such a long period, and the possible
long-term consequences of not doing this seem rather remote and unlikely. Actually, if his pet did turn out to have rabies, the owner
would probably be the person most likely to be bitten, but the law is not
only trying to protect the individual pet owner, but rather the entire
British society.

Not only does the pet owner, like the cigarette smoker, face the
same conflict between desirable and undesirable consequences from the
same activity, but unlike the smoker, the undesirable consequences
affect not only the pet owner, but also everyone else in the society. It
is difficult for the pet owner to conceive that smuggling in the pet kit-
ten could have such far reaching consequences for so many people. So
in this and other similar situations, the individual has difficulty changing his behaviour not only because the immediate consequences are
felt more strongly than the long term ones, but also because the individual finds it very difficult to imagine the enormity of the undesirable
consequences for the entire society which his actions could precipitate
in the future. The individual naturally feels the immediate and personal consequences of an activity much more strongly than he does the
long-term consequences for the rather abstract notion of the "good of
his society".

The distinction between these two factors which make it difficult for
people to change their behaviour even when they accept the validity of
a given hypothesis, is probably a bit arbitrary and artificial. This is
the case first of all because whether or not the consequences of a particular action concern only the individual or all of the society, they
always occur *in the distant future*. The threat of rabies in Britain
would occur several *days or weeks* after an infected animal was brought
into the country. The difficulty the individual finds in imagining the
enormity of the undesirable consequences for the entire society following a particular activity is actually only an additional factor which compounds the difficulity the individual has because the immediate effects
of an activity are felt much more strongly than the separated effects.
Whether the consequences of an activity effect only the individual or all
of society, these consequences are always in the distant future.

The second reason the distinction between these two factors is rather arbitrary and artificial is because each individual is also of course

part of the society. The person who smokes himself to death is not
only killing himself, but also part of the society. Undoubtedly more
people already die of lung cancer caused by smoking each year in Britain than would die of rabies after it was introduced. Enough people
smoking can cause just as much damage to a society as a contagious
disease. When an individual harms himself, he also, to a small extent, harms his society. An individuals's actions have consequences
not only for himself, but also for the society in which he lives.

Of course societies found long ago that people sometimes have great
difficulty in changing their behaviour in the desired manner even when
they have learned and accepted the long-term survival benefits of doing
so. All societies have attempted to solve this problem by instituting
various rules or social norms such as morals, taboos, and laws. By
instituting these rules, societies have attempted to overcome the problems caused by the conflict between immediate and long-term consequences of the same activity. All the norms in a society have a certain
punishment for misbehaviours ranging from mere social ostracism to
supernatural sanctions to physical force from the society. They often
also give various rewards such as social approval, supernatural benefits and honours for "correct" behaviour.

Without any rule or social norm, the individual will in certain situations have difficulty changing his behaviour because of the contrast
between the contiguous and separated consequences of the same activity.
The immediate consequences are felt much more strongly than the long-
term ones. Societies attempt to rectify this problem by introducing
yet another or *auxiliary consequence* for this same activity. Societies
introduce a third and hopefully immediate or near immediate consequence. If people have trouble changing their behaviour because the
undesirable consequences seem far removed in time and person, society attempts to help them along by introducing a new undesirable but
more immediate and personal consequence. Britain has attempted to
help pet owners change their behaviour in a direction favourable for
their own survival and that of the British society. If the long-term
undesirable consequences of smuggling a pet into Britain (the possibility
of spreading rabies to Britain) seem too remote and abstract for the
pet owner, the British government has attempted to help the pet owner
by instituting *other* undesirable consequences of a more immediate and
personal nature (i. e. a heavy fine or jail). Societies attempt to correct
the deficiency of attention which most people give the long-term consequences of an activity. Societies institute a new and hopefully immediate consequence which attempts to counteract the natural immediate

consequence. Norms are instituted to help people behave in a way
which they already accept is in their own and society's best interest.

It is quite possible then that all norms or rules are the result of
this conflict between the immediate or personal and the long-term consequences of an activity. For example, to my knowledge, no society
has ever found it necessary to institute a norm which requires people
to drink plenty of liquids on a hot day. The reason being that both the
immediate and long-term consequences of this activity are functionally
desirable. The immediate consequences of this activity to the individual (pleasure) is not harmful to the survival of society and the long-
term consequence (prevention of dehydration) is useful to the individual and the society. No norm is needed because there is no conflict
between the immediate and the long-term consequences. No other animal has norms or rules because they are not capable of anticipating
what may happen in the long-term future.

The type and amount of punishment or reward used as an auxiliary
consequence by a society for a particular behaviour is related to at
least three factors. First is the seriousness of the proposed long-
term consequences of this behaviour. No doubt the punishment for
smuggling pets into Britain would be much stronger if we did not possess a series of vaccinations which can be given to people who have
been recently bitten by a rabid animal. Without these vaccinations the
seriousness of an individual introducing rabies into Britain would be
much greater and therefore British society would no doubt have much
stronger punishments for violation of this norm. The more important
the long-term consequences of a behaviour are deemed to be, the more
society will attempt to ensure "correct" behaviour by instituting more
powerful rewards and punishments.

The second factor is the amount of immediate pleasure (or displeas-
ure) which this relevant behaviour creates in the individual. The
stronger the immediate natural consequence of the behaviour, the
stronger the immediate artificial or auxiliary consequence society will
introduce in an attempt to counteract this. Society must fight fire with
(an opposite but equal) fire. It is no good issuing a small fine to a man
who tried to rob a bank. A society's artificial immediate consequence
must outweigh the natural immediate consequence.

Thirdly, at least in modern societies, there is a tendency for people
to recognize the speculative nature of their hypotheses and consequently
to vary the amount of reward or punishment for violation of a norm to
some extent, depending upon how solid the empirical evidence is supporting the norm's underlying hypothesis. Modern societies tend to

use stronger rewards and punishments for norms which are based on hypotheses with solid empirical support than for norms based on hypotheses with skimpy empirical support. The more sure a modern society is of the validity of a given hypothesis, the more justified they feel in backing the hypothesis with auxiliary or artifical consequences.

For example, because the evidence supporting the link between cigarette smoking and lung cancer is not conclusive, societies have been somewhat reluctant to impose severe restrictions on cigarette smoking. Governments have taken some steps to curb cigarette consumption, such as higher taxation, restrictions on advertising and required health warnings but these steps have only attempted to make it more difficult to smoke, not to make it impossible. The severity of these regulations appears to be related at least in part to the amount of empirical evidence at hand supporting the cancer hypothesis. As more and more scientific evidence has accumulated supporting this hypothesis, modern societies have slowly increased regulations aimed at attempting to reduce cigarette consumption. The cigarette-cancer hypothesis proposes that a great deal of damage is being done to individuals (and hence also to their societies) by cigarette smoking. The severity of the consequences of this hypothesis makes it seem most likely that if new scientific evidence continues to support this hypothesis, it will only be a matter of time before many if not all countries ban the sale and consumption of cigarettes.

Thus the amount of reward or punishment a society uses to back up its norms is related to how important the long-term consequences of this behaviour are believed to be, how strong the immediate natural consequences of this behaviour are which the auxiliary consequences must overcome, and how sure a society is of the validity of the underlying hypothesis. Behaviours which are not deemed to be of great importance are usually only a "custom" or "manner" type of norm. Only social disapproval meets the misbehaver and social approval is given to the person who "behaves". Behaviours which are believed to be of great importance are upgraded to morals, taboos and/or laws.

Obviously the chances of "getting caught" and the immediacy of the punishment have something to do with the effectiveness of a society's laws. In order for taboos or morals to be effective of course, the potential wrong doer must accept the validity of his society's magical or religious beliefs. It is only when the individual *believes* that he will be punished by forces or spirits that morals or taboos have any teeth. Although the proposed auxiliary consequences for breaking a moral or taboo are of course not immediate (the proposed punishment in a per-

son's "afterlife" is the ultimate in delayed punishment), to the strong
believer, this proposed punishment has an immediate subjective effect.
The person's "guilty conscience" or fear of the spirits or diseases or
such is in itself an immediate punishment. Since the effectiveness of
morals and taboos depends upon the members of that society believing
implicitly that the proposed punishment will be carried out, as societies
become more secular, it is necessary for modern societies to rely
more and more on laws and less on morals or taboos for governing the
behaviour of their members.

   This process of adding a new and immediate auxiliary consequence
for a behaviour by a society will be referred to as the "legitimization"
process. A society "legitimizes" the reason for behaving in the desired way. The government and/or religious leaders establish a norm
with relevant rewards and punishments to accommodate the "correct"
behaviour based on the hypothesis accepted by that society.

   Once a norm has been legitimized (i. e. an auxiliary consequence
instituted), this activity is then labelled as either "good" or "bad" depending on whether the behaviour is being encouraged or discouraged
by that society. In an attempt to ensure the desired behaviour, society
institutes value or ethical connotations to the relevant activity. A behaviour which is deemed undesirable is referred to as "wrong", "evil",
"bad","sinful" and so on. Similar positive labels are given to desirable
behaviours. What is considered "good" and "bad" differs from society
to society, but "temptation" in all cultural contexts is when an individual considers satisfying his immediate and personal desires rather than
resisting in favour of what is believed to be good for society in the long
term. The important point here is that these ethical and value labels
are added only after the society has accepted a given hypothesis that
this behaviour has certain long-term consequences. What is deemed
"good" or "bad" in different societies varies because they accept different underlying hypotheses.

THE FORGOTTEN FUNCTION
   Probably the true functional nature of a norm is most clearly evident to the people of a society just before the norm is legitimized.
Once the members of a society have learned and accepted the validity
of a given hypothesis but many of the people nevertheless continue to
behave in a way which is believed to not be in the best interest of society, there is usually a great deal of talking and discussion by the society's leaders and others that the present situation is undesirable, and
to use a phrase often heard in modern societies, "there ought to be a

law" to correct the situation. The civil and religious leaders recognize that many people are unable to change their behaviour in the direction indicated by the accepted hypothesis. The violators are critized with such phrases as "what if everyone did that?" Meaning of course that the long-term survival of the society would be jeopardized if all the people behaved in this way. It is at this point, no doubt, that the largest number of people clearly recognize the functional reason and thus the need for legitimizing the norm. They accept the need by society to use force or supernatural sanctions as a necessary measure to coerce people to behave in a way which is believed to be useful or functional for the future of that society.

However, soon after a norm has been legitimized, the reason for legalizing or moralizing the norm is often forgotten by many of the people. The long-term consequences of a behaviour are remote and often rather abstract and therefore easily forgotten. Soon after the norm has been legitimized, many of the people in the society begin to forget the long-term survival value of the activity and hence the reason for the legitimization.

Once most of the people have forgotten the actual reason for legitimization of the norm, they often tend to think the behaviour is bad or wrong simply because it is "against the law" or because "it is immoral". Once a norm has become a law or moral, it is often accepted by most people as valid because of the auxiliary consequences. Often legitimization eventually becomes, to some extent, its own justification. Ask any automobile driver why he always stops at a red light and most people will answer without thinking "because it is against the law not to stop". Only if the person is pressed as to why the law exists in the first place will he add that this law is very helpful in preventing accidents and deaths at intersections. Ask a Christian person why adultery is a sin and almost without exception they will answer "because the Bible says so". In other words, it is a sin because the Bible says it is a sin. Only rarely would a religious person add that this norm against adultery is very useful in preventing the breakup of a stable family structure necessary for the rearing of children. Once the original functional reason for legitimizing the norm has been forgotten, many people accept the norm as valid simply because of the auxiliary reasons.

This problem of forgetting the functional reasons for legitimizing a norm is undoubtedly more acute for a society's morals or taboos than it is for the laws. This is because morals and taboos, unlike laws, are never believed to be created by humans. A taboo is believed to be a rule based on the nature of the world (e.g. stealing fruit inevitably results in hideous diseases) and a moral is supposed to be a rule issued by supernatural spirits (e.g. "Thou shall not kill"). The nature of magical and religious belief makes it very difficult for believers to admit to themselves that their morals or taboos are actually of a human origin rather than of a supernatural origin.

Remember that magical and religious rules have little effectiveness unless the people believe implicitly in the inevitability of the proposed punishments. A moral or taboo is effective in changing human behaviour only if the people accept completely that the proposed punishment will actually occur. Unlike a society's laws, there is no actual physical punishment to back up morals and taboos. Consequently, the realization of the long-term survival benefits of morals and taboos are often not only simply forgotten, but this knowledge is often actively and vigorously denied by the believers and their magical and religious leaders.

To admit that a moral or taboo had a long-term functional nature would open the door to recognizing the possibility that the rule had been created by humans in an attempt to improve conditions and was not, after all, commanded by the spirits with supernatural sanctions. Recognizing that a moral had a functional nature would imply that this rule might not have been sent down by the spirits and hence might not have any supernatural punishments. Recognizing the functional nature of morals and taboos would tend to destroy their effectiveness. Once people doubted the existence of the supernatural punishments, the norms would no longer have any teeth.

This is not to say that the religious or magical leaders who proposed these rules were necessarily conscious deceivers. They too accepted the belief system of their society and yet they realized that certain changes in the people's behaviour should be encouraged. They most likely convinced themselves that these rules were not of their own making, but had actually been inspired by the relevant spirits. No doubt Moses actually believed that the Ten Commandments had been sent to him from a divine source, but he also may have dimly recognized that only if his people firmly believed that these norms had a divine inspiration would they have any real effectiveness.

It is now possible to propose an answer to a question posed at the end of an earlier section. The question was how is it possible for a norm based on a magical or religious hypothesis to be nevertheless objectively functional for reasons unknown to most of the people in that society? How could the Hindus' prohibition on killing cows be based on a religious explanation ("cows are sacred"), but nevertheless be objectively functional for other reasons unknown by the people?

The answer I am suggesting is that at some point in the past at least some Hindu people realized the damage done to their society by farmers killing their cows during severe droughts, but that after this rule had been proposed and legitimized the recollection of the functional nature of this moral was not just forgotten, but actually suppressed by the Hindu religion. At some point many Trobrianders must have realized that the stealing of fruit from outlying trees was detrimental to the production of fruit, and when someone proposed that stealing fruit resulted in hideous diseases, this hypothesis was probably readily accepted by most people because they realized on some level that this proposal would reduce thievery. But after the legitimization of this taboo, its functional nature was quietly but firmly ignored and even denied.

Societies easily forget the functional nature of their laws because it is easy to do so, and they suppress the functional nature of their morals and taboos because it is essential to do so to ensure the effectiveness of these types of norms. It is possible for an objectively functional norm to be based on a magical or religious explanation because when the norm was first proposed it was based on a valid hypothesis (e.g. killing cows is detrimental because only cows can produce more oxen when the drought is over), but in the legitimization of this norm another or auxiliary explanation was added to ensure the rule was obeyed. The original hypothesis and function were forgotten and suppressed, leaving only the auxiliary magical or religious explanation.

CHANGES IN NORMS

All societies have rules governing the behaviour of their members, but the rules imposed by one society may differ greatly from those used in another society. It has been proposed that the norms in all societies have been based on hypotheses which predict future events. All norms encourage behaviours which are believed to enhance that society's survival in the future, and they discourage behaviours which are deemed not to have survival advantages. This section examines how a society changes its norms. What are the underlying processes and obstacles which a society must go through in order to change from one norm to another norm?

The primary factor causing a society to change one or more of its norms occurs when the society switches from accepting one hypothesis to accepting another. When a society has a "belief revolution" (discussed in Chapter 8) and abandons one explanation in favour of another, this often has consequences not only concerning their belief system and their ability to control the environment, but also on their norms. If a society changed from accepting a magical or religious explanation to accepting a scientific one, this change has far reaching consequences for: (a) their level of scientific knowledge, (b) their ability to control their environment. This might well result in an improvement in (c) their economic or health standards, that is their survival ability, and (d) the rules or norms which that society imposes upon its members.

For example, consider the consequences of Dr. Reed's discovery that yellow fever is caused by bites from infected mosquitoes (discussed in Chapter 5). After the scientific community had accepted Reed's hypothesis, this resulted not only in an improvement in (a) medical knowledge (b) our ability to prevent outbreaks of yellow fever (c) a reduction in deaths caused by yellow fever, but also (d) a change in the health regulations.

Before Reed, western societies accepted that yellow fever was caused by "fomites" that is germs which lie dormant in baggage, clothing, merchandise or furniture and were thus capable of conveying the spread of the disease over great distances. Since the validity of the fomite hypothesis was universally accepted, western societies enacted quarantine regulations which required the health authorities to "disinfect" baggage and clothing coming from ports where yellow fever prevailed. For example, the United States Quarantine Regulations of 1889 covered the following subjects: disinfection of houses; disinfection of premises; disinfection of freight; inspection of trains; regulation of traffic; disinfection of mail; inspection of mail; inspection of vessels; disinfection of vessels; inspection of territory and so on.

These rules requiring health authorities to "disinfect" and inspect all of these items now seems a bit silly to us, but they were perfectly reasonable if you accepted the validity of the fomite hypothesis. These rules were attempting to prevent outbreaks of yellow fever based on the hypothesis accepted at that time. The citizens believed these norms to be objectively functional. However, if an anthropologist from a much more advanced society had visited the United States at this time, he would have no doubt labelled this practice as "magic".

The publication and acceptance of Reed's hypothesis resulted in a change of norms. The old quarantine regulations requiring health authorities to "disinfect" baggage and clothing were changed to entirely new regulations which were in line with Reed's mosquito hypothesis. Now yellow fever patients themselves were quarantined with their rooms protected with wire screens (to prevent more mosquitoes from

becoming infected by sucking the blood of the patient). Also during a
yellow fever outbreak, oil was poured into all cisterns, barrels, pools,
ponds and other such receptacles of stagnant water, thus making it
impossible for the mosquito larvae to mature.

Another similar example occurred when western societies finally
accepted the germ theory, which proposes that infections are always
caused by germs entering an open wound. Although only Hungary accepted Semmelweis' ideas and in 1855 legitimized a rule requiring hospital authorities to follow his procedures, other men such as Pasteur,
Koch and Lister were eventually able to convince all western societies
of the validity of these proposals. With the acceptance of this hypothesis, treatment of wounds in hospitals and even homes changed drastically. Cleanliness quickly became the norm after the new hypothesis
had been accepted. Today in every country in the world, various forms
of laws and required health regulations demand that hospital staff maintain certain standards of sanitation and hygiene.

More recent examples include the gradual liberalization of the laws
in many countries prohibiting the use of marijuana with the acceptance
of scientific findings that marijuana is not, after all, seriously harmful
to humans. Also a law banning cyclamates was passed in the U.S.A.
in 1970 because of a hypothesis (supported by empirical evidence) that
cyclamates cause cancer in rats (and thus might also cause cancer in
humans). This law may now be changed again because of a new finding
which contradicts the earlier findings.

Once a society has accepted a new hypothesis, this often results in
a change of norms. If at some time in the future someone proposes--
and his society accepts--that blue-eyed, left-handed people who eat
carrots during a full moon will die in a few years of a heart attack,
then it will only be a matter of time before some type of norm results
which attempts to discourage this behaviour by these people. Of course,
societies do not change from accepting one hypothesis to another only
after a scientific discovery. A society can switch, for example, from
accepting a magical hypothesis to accepting a religious explanation.
These changes are usually related to influence from another society or
influence from a dynamic leader or such, rather than in response to
empirical evidence.

The second factor which causes a society to change its norms results from a change in the environment. Even if a society does not
change any of its hypotheses, if changes occur in the environment, this
may well necessitate a change in the norms.

Webb (1951, p. 254-9) gives an interesting example of how the English common law on water rights changed when it was transplanted to a different environment in America. The English common law of "riparian" water rights evolved in humid England and was designed to serve a society which had an abundance of moisture and running water. This law gives only the owner of the bank of the stream any rights to the water. A farmer who did not own land next to the stream had no such rights. If a farmer diverted water from the stream (to turn a mill wheel or such), he was required to return the water to the stream. Thus each owner of the bank had a right to the full and undiminished flow of the stream.

This law obviously precludes any possibility of irrigation, as water diverted for irrigation is absorbed and cannot be returned to the stream. In England this law worked well because irrigation was unnecessary. All the farmers could grow their crops (and thus survive) without having to use water from the streams for irrigation.

However, when this law was transplanted by English settlers in America, the change in the environment required a change in the law. The law worked well enough in the eastern states where the precipitation was similar to that of England, but as new states were admitted to the union in the drier plains, demands came from farmers that a modification of the law was necessary. In order to survive in these semi-arid regions, farmers had to irrigate their crops. The law on water rights was changed in these states to accommodate the change in environment. Webb shows that the amount of change in the law in each state varied with the amount of precipitation in that state. The most radical changes occurred in the driest states.

A more recent example occurred in 1973 when, in many countries, there was a temporary and artificial shortage of petroleum due to an embargo by many of the oil producing countries. Owing to the sudden shortage of petroleum (a change in the environment), many countries changed their laws relating to highway speed limits. Before this, speed limits were regulated only by the concern for safety, that is attempting to reduce traffic accidents. However, with the shortage of petroleum, many countries changed their laws concerning speed limits because driving slower uses less fuel. Speed limits were lowered because this conserves fuel. A new norm was required because the external environment had changed.

The line separating these two factors which result in a change in norms is sometimes a bit arbitrary. Often changes in the environment are a result of human activities. Often scientific discoveries (i.e. a new hypothesis) result in a change in the environment which then requires changes in the norms. For example, the invention of the internal combustion engine (which was the result of a hypothesis proposing how the energy stores in petroleum could be converted into mechanical energy), resulted in changes in the environment (e. g. the sudden appearance of a large number of automobiles). With the appearance of these automobiles, societies had to make numerous changes in not only the roads, but also create many new norms such as speed limits, requiring driving licences and safety inspection of the vehicles. New hypotheses often result in new environmental conditions which then require new norms. Modern societies are today passing new laws attempting to reduce pollution which is, in a large part, yet another change in the environment caused by this same discovery. A new hypothesis often has far-reaching consequences for the society which were completely unforeseeable to the person who first proposed the hypothesis.

A society legitimizes norms in an attempt to ensure survival of its members and itself in the future. Societies change their norms when the leaders and people either reject one hypothesis and then accept another, or when environmental conditions in that society change (either naturally or else as a result of human activities). Thus most, if not all, of the wide diversity in norms in different societies can be understood as a result of these two factors. Even though all norms in all societies are believed to be functional by the people concerned, the great differences in norms from one culture to another can be explained by the wide differences in the hypotheses accepted as valid in each society and the differing environments.

If this wide diversity in norms is a result of these two factors, then it seems most likely that in the future a large portion of this diversity will vanish. Since all primitive societies are gradually accepting our scientific hypotheses concerning medicine, growing of crops, and so on, the diversity of hypotheses explaining any one event, and therefore of the subsequent norms, is decreasing. There are thousands of possible magical and religious explanations concerning the cause of yellow fever, but there is only one scientific explanation (at least at any one time). As modern scientific knowledge is spread and accepted throughout the world, the number and variety of magical and religious explanations and their resultant norms must therefore be decreasing. Of course, the environmental factors will always remain, producing some diversity in norms.

When a primitive society accepts our superior scientific knowledge, this has far-reaching consequences and ramifications for their culture

which even we do not yet fully understand. Not only does this change
obvious aspects of their culture such as their level of scientific knowledge and ability to survive, but it also has less obvious but nevertheless
extremely important consequences for such things as their magical or
religious belief system, the prestige of their magical or religious leaders, their economic system and their values, rules and norms. It is
not possible to change the hypotheses accepted by a society without also
creating many other changes in that society.

CULTURAL LAG

   After a society has changed one of its hypotheses or had a change in
its environment, there is an inevitable delay before there is a corresponding change in other parts of the culture. Ogburn (1950, 1957)
termed this delay "cultural lag". Societies must make certain adaptive
changes in response to a new hypothesis or a change in the environment,
but there is always a time lag before these changes can occur.

   One of the examples Ogburn uses is the changes which resulted in
societies after the invention of the automobile. Although the roads in
most countries in 1910 were sufficient for the first automobiles, as
cars were improved and were capable of greater speeds, societies had
to require improvements in the roads such as widening, improving the
surface, and eliminating sharp curves. However, it was several years
after the need for these changes was recognized before the changes
were carried out. This gap between the change in the environment and
society's reaction is an example of "cultural lag".

   There is also a cultural lag between a change in hypothesis or the
environment and the society changing the relevant social norms. It
was several years after the need became apparent that most societies
institutionalized the various rules necessary to curb the excesses of
the use of the automobile. As previously mentioned, such norms include speed limits, requiring driving licences, safety inspection of
vehicles and pollution controls.

   After a change in a hypothesis or a change in the environment, what
are the factors which encourage "cultural lag"? Of course, some delay
is inevitable due to several purely practical aspects such as the time
required to pass a law and so forth. But there often seems to be an
unduly long time lag before the changes in the norms are accomplished.
What are the factors which cause long delays in the legitimization of
new norms?

   Ogburn listed several such factors causing cultural lag, the most
important of which was what he called "vested interests". In any given

society, certain individuals have--either by inheritance, luck or superior ability--acquired a larger share of the power, prestige or wealth
than many of the other individuals. These powerful or wealthy individuals naturally attempt to protect or defend their advantageous position.
These "vested interests" oppose any change in society which would
undermine their privileged position. Consequently, whenever a new
hypothesis is proposed which would ultimately result in a change in
society which would be detrimental to their "vested interests", these
individuals oppose first the validity of the proposed hypothesis, and
later, after the hypothesis has been accepted by most people, they oppose the implementation of the resultant norms.

A good example here would be when the hypothesis "smoking causes
cancer" was introduced. The owners, managers and even employees
of tobacco farms and cigarette companies could in this case be considered the "vested interests". They all acquired their wealth and in
some cases prestige from the sale of tobacco to the public for smoking.
The cancer hypothesis threatened their vested interests. Consequently,
these people challenged the validity of this hypothesis (they even conducted some experiments) and now after the hypothesis has generally
been accepted by most people, they are still opposing the legitimization
of norms to discourage the smoking of tobacco.

Because powerful or wealthy individuals are often in positions of
power and authority, their opposition to changes in society is usually
far stronger than their small numbers merit. Even if the changes proposed in the society would benefit the vast majority of people, these
relatively few powerful individuals are often able to block the proposed
changes for long periods.

There is a second important factor which increases "cultural lag"
in the field of norms which Ogburn did not recognize, however. This
factor is related to the practice of legitimizing norms and the subsequent forgetting of the functional nature of these norms discussed earlier. After a norm has been legitimized the people often forget or even
suppress the original functional reason for institutionalizing this norm.
This phenomenon of forgetting or suppressing is undoubtedly more
acute for a society's morals or taboos than it is for the laws for reasons already explained.

When a new hypothesis is proposed or a change in the environment
occurs which suggests a change in that society's norms are needed,
there is invariably a lag before these changes are instituted. Part of
the reason for this lag is that many of the people in the society have
forgotten (or were never aware of) why the norm was originally legiti-

mized. The original functional nature of the norm may have long since been forgotten, and the individuals now alive have been taught since childhood that this behaviour should always be avoided simply because it is "bad" or "wrong" or "immoral" or "illegal" or such. Because the original functional reason for the norm has been forgotten, most of the people in a society may believe that the norm exists only because of the auxiliary consequences or reasons. For example, most of the Hindus alive now believe that one should not kill cows simply because "cows are sacred".

When a new hypothesis or a change in the environment occurs which suggests that one of the present norms is not functional, at least a few members of society will propose that the norm should be changed. The trouble is that many members of the society believe that the norm is justified only because of the auxiliary consequences. Since most of the people are not aware of the original functional reason for legitimizing the norm, they are not impressed by arguments that this function is no longer there. Cultural lag is often vastly increased by this fact that many people in society are unaware of the original functional nature of their morals, taboos and laws.

For example, if at some time in the future, India's reliance on oxen for ploughing the fields is eliminated, some Indians will probably pro-pose that it should no longer be immoral to kill cows. Since the func-tional purpose for not killing cows would no longer be in effect, the "progressives" would advocate elimination of the rule. However, the vast majority of Indians would probably reject this proposal. The majority would accept that it is wrong to kill cows simply because "cows are sacred". Arguments by the progressives that the conditions had changed would not convince the "conservative" majority.

A great deal of cultural lag is created because of this problem, es-pecially in the area of religion and morals. The reason is of course that the "progressives" and the "conservatives" are not really speaking the same language. The "progressives" are assuming that a law which is no longer functional, should be changed and the "conservatives" are assuming that their morals are dictated by the relevant spirits or forces and should remain the same regardless of changing circum-stances. Naturally there is very little understanding or real commu-nication between these two groups. Often reform of a norm is delayed many years because of this problem. The gap between the changed en-vironmental conditions and society's norms may become appallingly large before the "progressives" win enough converts to enact the needed reforms.

Societies must institute auxiliary consequences and ethical values
to ensure that individuals behave in a way which is believed to be helpful for the survival of that society in the future. However, the original
functional reason for legitimizing a rule is often forgotten or suppressed and as a consequence, when hypotheses or environmental conditions
change, there is a long lag before the relevant norm is altered. The
adding of auxiliary consequences and the forgetting of original functions
are a normal and probably necessary part of the legitimization process;
however these same processes sometimes result in very long delays in
reform.

If societies change their norms in response to (1) a change in a hypothesis accepted by society or (2) a change in the environment (and
often these environmental changes are created as a result of scientific
inventions), then it follows that a great deal of the change in norms in
modern societies is either a direct or indirect result of scientific advances. New hypotheses such as Reed's mosquito hypothesis of yellow
fever may result in direct, almost immediate changes in norms, whereas new scientific inventions such as the steam engine or internal combustion engine may eventually result in many changes to the environment which will require more and more changes in norms. The tremendous changes that have occurred in western societies since the
middle ages because of the "Industrial Revolution" were all of this
nature. The process of industrialization is actually possible only because of the many new hypotheses concerning the utilization of new
sources of power and improved techniques of mechanization. For example, it was never necessary to pass a new law outlawing unsafe
factory working conditions until after our scientific knowledge had
reached a level where it allowed us to build large factories.

If the rate of scientific discoveries has been increasing rapidly in
the past few centuries, then the rate of cultural change must also have
been increasing greatly recently. Ancient societies were apparently
able to function for centuries with very little change, whereas modern
societies must amend their laws every year. Modern societies' rapid
changes are caused by our rapid scientific advances.

Cultural lag has been defined as the time gap between a change of
hypotheses or environment and society's adaptation to these new conditions. If modern societies are changing more rapidly as a result of
more frequent scientific advances, then it seems likely that we have
more cultural lags than did the older, more stable societies. Since
ancient societies had a relatively slow cultural change, their cultural
lags must have been fewer and farther between.

Although modern societies must have many more cultural lags in a given time period, it seems most likely that these lags do not persist for as long as did lags in the older societies. This is because most norms in older societies took the form of morals or taboos rather than laws and consequently the problem of forgetting or suppressing of function after legitimization would have been much more severe and the "conservative" opposition to the changing of the norms would have been far greater.

However, as Ogburn pointed out, cultural lags in modern societies may pile up to an extent where the normally slow process of adaptation and reform of norms is no longer adequate to resolve these lags. In these circumstances, the stresses created by these lags may erupt into violent conflict between the "progressives" and "conservatives". This is the process we normally refer to as "revolution". A revolution could be defined as an attempt to clear away at once all the accumulated cultural lags.

EDUCATION

All human societies accept hypotheses proposing connections between events separated by time. All societies also have norms or rules which attempt to direct and restrict the behaviour of their members. I have proposed that the norms in each society are based upon the hypotheses which are accepted by that society. Norms attempt to encourage behaviour which is believed to have survival value for that society in the future and to discourage behaviour which will not help survival in the future.

All societies, primitive and advanced, also engage in some type of educational process, primarily for the young. In most primitive societies this education is largely an informal and unconscious process. It is only in modern societies that specific institutions such as colleges and universities for the training of the young have been established. However, all societies have some type of educational system for the gradual imparting of knowledge, values, skills, techniques, manners and customs to the young.

This "educational process" can be viewed as simply a cluster of norms in which societies attempt to ensure the survival of their members in the future. Education is almost always accepted as having a functional purpose: that is, of preparing the young with skills and knowledge which they will need in the distant future. "Education" is simply another instance where societies are attempting to prepare for what they believe (on the basis of hypotheses) will occur in the distant future.

All education of the young involves at least one basic hypothesis: that these young people will grow up and become adults and face conditions in which the possession of certain skills will be advantageous for their survival. This "growing-up" hypothesis is actually only part of the "birth-death" hypothesis discussed in Chapter 7. The idea that all children will become adults and then eventually die is actually only a hypothesis.

Also, societies usually assume that skills and knowledge which their children will need in the future will not be too dissimilar from the skills needed in the present society. In particular, the skills which societies require their young to learn depend upon (1) the type of skills and knowledge needed in the present society and sometimes (2) predictions of how the future may differ from the present. For example, in hunting and gathering societies the boys learn skills such as how to hunt or fish and how to take care of themselves in the wilderness and the girls are taught appropriate skills such as gathering, tanning and making clothing from skins. In farming societies adults teach the young knowledge such as sowing, ploughing and pottery making. Unlike some norms, the functional nature of education is almost always at least partially recognized.

As with other types of norms however, cultural lags sometimes develop. Whenever a society undergoes a rapid transition, it is quite likely that at least for a while the young will continue to be taught skills which, in fact, will not be of any use to them in the future. A hunting society which had just settled down and begun farming, might very well continue to teach the boys hunting skills for a period. European societies continued to require that some of their children learn Latin long after this language had ceased to be an important language in academic and scientific discourse.

Also, if a society accepts certain hypotheses which are, in fact, invalid and therefore defined as "magic", this could very well mean that they require their children to spend a great deal of time and energy learning skills and "knowledge" which do not, of course, have any objective functional survival value. Obvious examples include learning of many of the witchdoctor's skills and "medicines", the various methods of fortunetelling and divination, as well as perhaps many years of theological study required for modern priests and preachers. But since all peoples (including ourselves) believe that all of the hypotheses they accept are valid or scientific, all education is naturally believed by each group to be objectively functional. Undoubtedly some of the

hypotheses which we accept as valid will someday be recognized as invalid or "magic". Part of the education in any society is undoubtedly
useless for this reason, the problem consists of identifying the useless
or non-functional parts.

Educational norms, similar to other norms, are necessary because
of the conflict between the immediate and long-term consequences of
the same activity. Studying and practising may have long-term survival benefits, but the immediate consequences of these activities for the
student are often hard work and boredom. Consequently, societies
usually add more immediate auxiliary consequences and values to encourage the student to continue with this useful behaviour. A system
of rewards and punishments (e.g. "good" grades, praise, honours, and
ridicule, withdrawal of affection and even physical pain) are usually
used to encourage the young to continue this activity when their natural
curiosity and desire to learn and imitate their elders wears thin.

Recently a technique known as "behaviour modification" has become
a popular method of teaching and therapy. Behaviour modification is
simply a more sophisticated and refined method of encouraging or dis-
couraging (by the use of contiguous rewards and punishments) certain
behaviours which the teacher believes (on the basis of a separated hypothesis) will be useful to that individual in the distant future. Behaviour modification, like other forms of education, uses contiguous consequences to reinforce behaviour which, as a result of separated
learning, the teacher believes will have favourable long-term consequences for that individual. The contiguous rewards and punishments
are the auxiliary consequences which the teacher adds to counteract
the natural immediate consequences.

Humans are the only animals which purposefully teach their young
skills which will be needed in the distant future because we are the only
animals which predict this long-term future. Although some animal
parents sometimes seem to be "setting an example" for their offspring,
it is doubtful if this could be called purposeful teaching. For example
Kuo (1930) allowed some kittens to see their mothers killing mice,
while the other kittens did not. Most of the first group of kittens
became mice killers, but only about half of the second group did. However, since the behaviour of the mother would probably have been the
same whether or not the kittens were present, this example should not
be counted as teaching in the sense that the parent was not specifically
attempting to prepare the young for the future. Certainly some animals
are capable of imitation, that is learning by watching other animals
(see p. 80-81), but it would be unwarranted to conclude that the first

animal was *teaching* the second animal.

In other examples, animal mothers seem to be "encouraging" certain behaviours in their young. For example, mother apes seem to encourage their young to walk. However, except for the well-known and common maternal rebuffing of their young when they begin to reach the age of maturity, punishment seems to play no part in other animals' "education". Certainly, the "rebuffing" behaviour and also possibly the "encouraging" behaviour could very well be instinctive (and this could be tested). The lack of punishment in other animals, strongly indicates that the "teaching" animal is not knowingly attempting to impart knowledge to the young. The fact that adult animals do not punish their young when they do not learn, certainly points to the conclusion that the adults are not consciously trying to teach them.

It is likely that some form of education began soon after the first humans evolved. Education, like so many other aspects of human activity such as religion, science, language, art, magic, law, morals and taboos, is intimately related to, and a result of, our unique ability of "imagination". Education must have begun soon after humans began hypothesizing and therefore predicting the distant future. Education is just one of the many ways in which we attempt to prepare for the future.

CONCLUSIONS

There is still one question which has not yet been fully answered in this chapter. We have still not concluded whether all norms are actually functional. Is it possible that the many differing and conflicting norms in different societies are all functional?

So far it has been concluded that norms differ from one society to another because of (a) differing underlying hypotheses (magical, religious and scientific) accepted by each society and (b) differing environmental conditions. From this proposition it follows at least that all norms are *believed* to be objectively functional (i.e. help control the environment in the future) by most of the members of each society.

Leaving aside for the moment the question of whether all norms are actually objectively functional, it follows that since all peoples *believe* their hypotheses and the resultant norms to be valid and objectively functional, that therefore all norms must at least be *subjectively* functional (i.e. provide emotional comfort). Since everyone believes their norms to be objectively functional, these norms must at least be subjectively functional. At the very least, all norms are subjectively functional.

Now the more difficult question arises: is it also possible that all norms are objectively functional? Since some norms are based on magical and religious explanations, it would at first seem to follow that these norms must be objectively non- or dysfunctional. However, as discussed earlier, it is sometimes possible for such a norm to be objectively functional for *other* reasons which are unknown to most of the people in that society. I have proposed that this amazing phenomenon occurs because of the process of legitimization (i. e. the adding of auxiliary consequences) and the subsequent forgetting or even suppression by society of the original functional reasons for instituting this norm.

It seems most unlikely however, that all norms based on magical or religious explanations are actually objectively functional for hidden reasons. No one has proposed for example, that members of the Zande tribe who put medicine juice on their plants to make them grow better are actually engaging in an objectively functional activity. (Some anthropologists have proposed that this activity has a subjective function, but no one has suggested that it has an objective function for other hidden reasons). It is most difficult to imagine what unknown objective function this activity might have, although of course this possibility always exists. Similarly, it is most difficult to conjure up what unknown objective function the practice of bloodletting by doctors in Europe in the middle ages might have had. Bloodletting may well have had a subjective function, but it is difficult to see what hidden objective benefit may have accrued from this practice.

Thus it follows that a particular norm may be subjectively functional, but objectively dysfunctional! It is possible that by encouraging a particular behaviour, a society may gain some emotional benefit, but nevertheless bring death and destruction upon themselves in the future. It seems likely that doctors and patients both received some subjective emotional comfort from the practice of bloodletting (by believing that they had some control over diseases) but the practice nevertheless killed many patients! Similarly it is even conceivable that a society could institute a norm which would ultimately result in their own extermination, but this norm would, nevertheless, give them a certain amount of emotional comfort as they unknowingly plotted their own destruction.

Since all norms are subjectively functional, the more important consideration should be, is the norm objectively functional (for either the reasons stated by that society or for other hidden reasons)? Since norms are instituted in an attempt to control the environment in the future, the real test of whether a norm is effective or not should be whether it is *objectively* functional. Subjective function is important to prevent emotional despair in the face of terrifying dangers, but by itself, it is only a poor substitute for objective function.

It is also possible that a norm may be temporarily objectively dysfunctional due to changing environmental conditions and cultural lag. Due to the delay caused by cultural lag, a formerly functional norm may continue to be enforced for a time even though it is now dyfunctional.

Therefore we may conclude that although all norms are subjectively functional, the same can not be said for objective function. It is possible for norms to be temporarily dysfunctional (due to cultural lag) or permanently dysfunctional (due to an invalid underlying hypothesis). Although some norms based on magic or religious explanations may be functional for other hidden reasons, it seems most unlikely that this is always the case.

The ideas proposed in this chapter are, at least to some extent, testable. The basic proposals of norms based on hypotheses, legitimization by society, the eventual forgetting of the original function and cultural lag after a change in hypotheses or environmental conditions are processes which may be observed and even partially experimentally tested by scientists.

Perhaps the simplest method of empirically supporting these ideas would be by naturalistic observation of a primitive society or even modern societies after a change in an accepted hypothesis or a change in the environment had occurred which had important consequences for the survival of that society. In primitive societies, anthropologists could observe how a society's norms change as a result of that society gradually rejecting its magical or religious explanations of such things as illness or poor crops and its acceptance of modern scientific hypotheses. In modern societies, social scientists could simply observe what happens in our legislatures after an important scientific discovery has been made which affects our future survival. For example, if some day someone proposes (and society eventually accepts) that lung cancer is not caused by cigarette smoking, but rather by watching too much television, social scientists could observe how the acceptance of this new hypothesis would eventually effect our laws and norms concerning cigarettes and television.

It would also be possible to test these proposals in an actual experimental setting. Sherif (1956) did some interesting experiments on group conflict in a summer camp for boys. The boys, not aware that they were part of an experiment, were divided into two groups and then Sherif introduced either conflicting aims (when one group could achieve its goal only at the expense of the other) or common aims. He was then able to observe how these conditions affected the relationships between the groups, between members within each group and between members of the different groups.

In a somewhat similar manner, it would be possible to test some of the hypotheses advanced in this chapter. In a small isolated community (such as a summer camp) experimenters could artificially introduce a problem (i.e. a change in the environment) such as a shortage of water due to a "fault" in the plumbing system. It would then be possible to observe how the group's norms were effected by this change. It would also be possible to introduce a hypothesis (which would probably be accepted by the group) which would propose a solution to this problem (such as having a plumber explain to the group that the fault was caused by excessive vibration in the pipes due to too much noise in the dining room!) It is unlikely however, that such small temporary groups (such as at a summer camp) would display the more slowly evolving phenomena of the forgetting of the original function and cultural lag.

In conclusion, it follows from the proposals presented in this chapter that the many differing rules enforced by different societies on their members are not just a haphazard jumble of restrictions arbitrarily forced on the people, but rather can be seen to be a lawful and logical system of attempts to ensure favourable conditions for each society's existence in the future. The rules differ between societies because of differing underlying hypotheses and environmental conditions in each society. With our superior knowledge we may now see that some norms enforced in our past or in today's primitive societies were objectively unnecessary or even counter productive. But the people who instituted these rules were acting in good faith in attempting to create a better society in the future. We should not be too critical of their mistakes as it is almost certain that some day some of our own laws and morals will also be found to have been unnecessary or even counter productive due to invalid underlying hypotheses.

All societies have used norms to direct the behaviour of their members, and it is likely that this will continue to be the case in the future. This is so because one of the most important aspects of human intelligence is attempting to anticipate what will happen in the distant future and norms are the inevitable result of this anticipation. As our ability to predict the future becomes more accurate however, our rules will change in light of this new knowledge. As we are better able to admit our ignorance, some of the unnecessary norms can be dropped, but on the other hand as our knowledge of separated relationships grows, the necessity of new norms will also become apparent.

No human society can ever be completely "free", that is without rules because it is part of the essence of being human to try to anticipate and then control the future. We can never be completely free to do what we want to do *now*, because we always imagine what consequences these actions may produce in the *future*. Our freedom is limited by our attempts to know and control the future.

Chapter 11

Sexual Norms

> In the beginning God created the heaven and the earth . . .
> And the Lord God took the man, and put him into the garden of
> Eden . . . and commanded the man, saying, of every tree of
> the garden thou mayest freely eat: but of the tree of the knowl-
> edge of good and evil, thou shalt not eat of it . . . for God doth
> know that in the day ye eat thereof, then your eyes shall be
> opened, and ye shall be as gods, knowing good and evil . . .
> The Lord God made he a woman, and brought her unto the man
> . . . and they were both naked, the man and his wife, and were
> not ashamed . . . When the woman saw that the tree *was* good
> for food, and that it *was* pleasant to the eyes, and a tree to be
> desired to make one wise, she took of the fruit thereof, and
> did eat, and gave also unto her husband with her; and he did eat.
> And the eyes of them both were opened, and they knew that they
> *were* naked; and they sewed fig leaves together, and made them-
> selves aprons . . . And the Lord God said, behold, the man is
> become as one of us, to know good and evil . . . therefore the
> Lord God sent him forth from the garden of Eden, to till the
> ground from whence he was taken.
>
> Genesis 1-3

The preceding chapter outlined the nature of norms or rules in human societies and how these norms change. Many of the examples
used in that chapter were from primitive or non-western societies.
The basic processes outlined of norms based on hypotheses, the legitimization (i.e. the adding of auxiliary consequences), forgetting of the
original function, a change of hypotheses or environment, and the resultant cultural lag, were probably relatively easy for the Western
reader to accept, especially as most of the examples used, at least in
relation to magic and religion, were from other societies. A Hindu
from India might well have strongly objected to my analysis of the origin and function of his norm against killing cows, but most other people would probably accept the logic and at least the possibility of its

validity.

It is usually a great deal more difficult for people to analyze critically their own society's beliefs and norms than it is for them to see
beneath those of other societies. Probably because a person usually
accepts the magical and religious explanations of his own society at
face value before he is capable of thinking critically, most people then
find it very difficult to go back and overturn beliefs which they have accepted for many years. A foreigner is thus sometimes able to perceive
the nature of a society much more clearly than are the natives.

The nature and functions of norms discussed in the previous chapter
are not only valid for primitive societies, but also for our own. Our
rules are also subject to these same processes. This chapter will
attempt to show that norms regulating premarital sexual behaviour in
our culture (as well as others) follow the same principles already proposed. Probably many more readers will object to the proposals in
this chapter because I am also analyzing our own morals instead of
only those from remote or primitive societies. Sexual behaviour in our
society has been subject to legal and moral restraints for as far back
as our records go, consequently many people have strong feelings
concerning this subject.

IGNORANCE OF THE SEX-PREGNANCY RELATIONSHIP

First it will be necessary to review a topic originally discussed in
Chapter 7 (p. 162-167). It was proposed that the knowledge that sex
causes pregnancy was the result of a pre-historic scientific discovery.
No other animal is aware that sex causes pregnancy because only humans are capable of learning a relationship between events separated
by time. At least several weeks must pass between sexual intercourse
and the first signs of pregnancy.

There is little evidence available at present to show when this pre-
historic discovery was first made. However, for reasons discussed in
p. 167-169, it seems likely that humans had learned the sex-pregnancy
relationship at least in relation to animals before they began using domesticated animals as a major source of their food. Present evidence
indicates that this first occurred somewhere around 10,000 years ago.

Anthropologists have found several primitive tribes that were either
completely ignorant of the sex-pregnancy relationship or else had only
a rudimentary understanding of this connection. A summary of Malinowski's of the people of the Trobriand Islands who were completely
ignorant of this relationship was given in Chapter 7. The Trobrianders
believed that women became pregnant only when a spirit (baloma) en-

tered into the womb of the woman. They believed that the child came exclusively from the mother. The whole process of reproduction lay with the spirits and the females, the men having nothing to do with it.

Anthropologists have also studied several other tribes in which the sex-pregnancy relationship is not clearly understood. Some of these tribes recognize some sort of vague connection between sex and pregnancy, but still have not learned that sexual intercourse *causes* conception. For example, Montagu (1974), after reviewing the literature on the various aboriginal tribes of Australia, concluded that most of the tribes were ignorant of this causal relationship, but that some of these tribes believed that intercourse was first necessary to prepare the woman for the entry of a spirit-child, although this preparation was not in itself the cause of pregnancy.

A little higher up on the scale of knowledge of this subject are the people of the Island of Mangaia. Marshall and Suggs (1971) found that although these people recognize that sex causes pregnancy, they believe that pregnancy results only from having intercourse repeatedly with one man.

Evidence from our own civilization's history suggests that our ancestors were also not very clear about the exact nature of this sex-pregnancy relationship. For example, Diodorus Siculus reported that the ancient Egyptians believed that only the father was a biological genitor to the child. They believed that the mother only provided a nidus and nourishment for the foetus. The afterbirth was regarded as the spiritual or physical double of the child.

All this evidence strongly indicates that at some point in our prehistory (perhaps as little as 10,000 years ago), all humans were completely ignorant of this connection. And even after the causal relationship was finally discovered, the exact nature of this relationship was probably not understood until relatively recently. It was only after a pre-historic scientific discovery had been made, that humans first acquired this very important knowledge.

MARRIAGE

In all societies, members of the opposite sex copulate at times and places approved of by that society. In the societies just discussed where the people were ignorant of the sex-pregnancy relationship, individuals engaged in sexual behaviour simply because it was a pleasurable activity. As proposed in Chapter 6, sexual intercourse is instinctive in all bisexual species.

The activities of men and women who engage in sexual behaviour on only a temporary basis, with no familial obligations or responsibilities imposed, might be referred to as "mating". Many societies provide for matings in one form or another, and no doubt matings occur in all societies whether they are sanctioned or not.

However, all societies provide for a second type of union between men and women. The institution of marriage exists in all societies. Needless to say, the form of marriage varies tremendously from one society to the next. For example, some societies allow only monogamy (the marriage of one man to one woman) whereas others also allow polygamy (one individual married to two or more spouses). Some societies require an elaborate ceremony before the individuals are pronounced "married", whereas in other societies no ceremony is required and the people are considered to be married simply when they begin sleeping together regularly.

"Marriage" in all societies has one aspect in common. Marriage always consists of a set of norms which encourage a long-term union between men and women in which the participants accept the responsibilities of rearing any children which are born. Often marriages are expected to produce children--in some societies, a marriage is not valid until the first child is born.

The rearing of children is a very important activity in all societies. The human infant needs care and protection by adults for a longer period than do the young of even the highest anthropoid apes. Human children need protection and help for at least 8 to 10 years even in the simplest societies. In more advanced societies, where longer and longer periods of education are required before the young person is capable of supporting himself, this period of dependency may last up to 20 or even 25 years.

Since the young in all human societies require this very long period of dependence on adults, all societies must cope with this important problem. Particularly in primitive societies, it is very difficult, if not completely impossible, for the mother to provide food, shelter, clothing, care and protection for her children for many years by herself. At least in primitive societies, it is necessary for at least two adults to work together in order to provide the necessary care and protection of children. In modern societies it has recently become possible, because of a greatly increased standard of living, free state day schooling and by hiring occasional baby-sitters, for one adult to provide this necessary care, on his or her own. Some of the consequences of this very recent development will be discussed later.

Because the "mating" type of union between men and women is, by

definition, only temporary, this type of relationship would not therefore be suitable for the rearing of children. If the couple were together
for only a short time, then only the mother would be available for the
rearing of the children. For this purpose, the society must encourage
the two (or more) people to remain together for a considerable period
of time. Marriage exists, and is functional in all cultures, because it
fulfills this universal need of two or more adults remaining together
for long periods and working to provide for the extended dependency of
human children.

Because the children in all societies have this long period of dependency before they can become self-sufficient, all societies encourage
the notion that children should be born only within the institution of
marriage. "Marriage" is a set of norms in all societies which attempts
to ensure the long-term care and protection of their newly arrived
members. Marriage is functional because it is attempting to ensure
the survival of that society in the distant future. Marriage is always
a licence to have children.

Even in societies which have not yet learned the sex-pregnancy relationship, girls are encouraged to have children only after they are
married. Among the Trobrianders for example, it is considered im-
proper for a woman to have illegitimate children. From our point of
view, this seems unfair because the single girls are allowed to have as
much sexual intercourse as they wish, but they are the object of scorn
and disapproval if they have children! However, they see no contradiction in this because they are not aware that sex causes pregnancy.

When Malinowski (1929) asked the natives why it was considered
wrong for unmarried girls to have babies, they replied "because there
is no father to the child, there is no man to take it in his arms".
(Remember that the term "father" among the Trobrianders is defined
only socially, that is, the man married to the mother). Thus, it is
considered wrong for unmarried girls to have children because there
is no man to "take the child in his arms" and to help her in nursing and
bringing it up. Illegitimate births are discouraged among the Trobrianders, as in all other societies, because the mother needs a defender
and provider to help her rear the children.

The Australian aborigines (Montagu, 1974, p. 257) also regard illegitimate births as a very shameful thing. They are also ignorant of the
sex-pregnancy relationship, and believe that pregnancy is caused by a
spirit-child entering a woman. Actually, few single girls get pregnant
in their society. We now know this is because of a combination of "adolescent sterility" (see p. 165) and early marriage age. The aborigines

themselves believe, however, that this is because "normally spirit-
children only enter married women". However, when an unmarried
girl does give birth, they conclude that such births are caused by someone, such as a medicine man, who wished to injure the girl's reputation
and has, by the use of magic, caused a spirit-child to enter her.

No other species of animal has rules or norms because no other
animal is able to predict what will happen in the distant future. Consequently, "marriage" does not exist in other animals for this same reason. No other species encourages, through the use of norms, members
of the opposite sex to remain paired together for long periods of time.
Many animals are promiscuous, that is they mate with several members of the opposite sex during mating season, thus even if the species
is "social", (that is the animals form groups, flocks, herds, etc.) the
father does not usually help in the rearing of his offspring. If the species is "non-social", (e.g. bears, orang-outangs, where individual
animals do not normally form groups except during mating season) the
mother must rear her young by herself. In a few species such as the
stickleback fish, it is the male who tends the offspring by himself and
the female leaves after laying the eggs.

There are however, some species of animals (especially birds)
which are monogamous and remain paired together for some period
after mating, with a few species, such as greylag geese, even "mating
for life". Humans easily have the longest period of dependence of any
animal before our young are able to become self-sufficient. However
in some other species, although the period of dependency for their
young is much shorter, it is absolutely essential for two adults to work
together to provide the necessary care and protection for their offspring. For example, among some species of birds, it is necessary
for one parent to sit on the eggs, keeping them warm and protecting
them while the other parent searches for food. It would be impossible
for one parent to do both of these jobs alone. Consequently in these
species, the two parents remain together for at least long enough to
rear the young, sometimes longer. However, this behaviour is instinctive. Among different flocks of a given species, the behaviour is
identical in this respect. Since the behaviour is controlled instinctively,
there is no room for group variation. Human societies, on the other
hand, differ a great deal concerning the details of marriage. It would
also be possible to conduct tests (see Chapter 6) to show that this behaviour in other animals is indeed instinctive and not learned.

PREMARITAL SEXUAL NORMS
All human societies restrict and direct the sexual behaviour of their

members in some way. No society is completely "free" in allowing
their members to engage in sexual activity at any time, at any place,
with any one they wish. Even in the most permissive of societies regarding sexual behaviour, there are some restrictions such as incest
norms, or restrictions on copulation during or after menstruation or
prohibitions on intercourse with individuals from certain clans and so
on to limit the freedom of the people.

These sexual norms differ widely from one society to the next. Although all societies restrict the sexual activity of their members in
some way, the nature and extent of these restrictions, particularly in
regard to premarital sex, varies tremendously. A review of the different norms concerning sex in different societies can be found in
Ford & Beach (1951, p. 180-192).

A minority of primitive societies studied by anthropologists were
found to have relatively strong prohibitions of sexual behaviour in
young people before marriage. These societies use various methods
of attempting to prevent premarital sexual activity such as threats of
severe disgrace or physical punishment, strict chaperonage of girls
and segregation of the sexes. For example, among the Hopi, an attempt is made to keep boys and girls apart from the age of ten until
marriage; the girls are accompanied by an older woman whenever they
go out. The most severe of punishments occurs in the Gilberts where,
if knowledge of premarital sex becomes public, both partners are put
to death!

Most of these restrictive primitive societies also attempt to prevent
the young from obtaining knowledge of sexual relations. In the western
Carolines the natives never discuss sex before children, especially
girls. Cuna children are kept completely ignorant of sexual knowledge
until the marriage ceremony. In these restrictive societies adults
often attempt to keep children in total ignorance of sexual intercourse
and the reproductive process. Chagga children are told that babies
come out of the forest.

In most of these societies, the boys are less carefully controlled
than the girls. Among the Hopi, the girls are expected to remain
chaste until marriage, but the boys are not similarly restricted. The
Hopi place all the blame for illegitimate pregnancy upon the girl. She
is scolded by her family and ignored by her friends, but the boy is not
regarded as at fault. Similar situations occur in other restrictive
societies such as the Kiwai Papuans of New Guinea. There are, of
course, exceptions to this such as the example given above where, in
the Gilberts, both partners are punished. However, the net objective

of sex norms in these restrictive societies seems to be attempting to
ensure the virginity of girls (but not necessarily boys) until marriage.
Some of these restrictive societies not only attempt to prevent premarital heterosexual activities, but also to prohibit other forms of sexual behaviour such as masturbation and homosexuality. Among the
Apinaye, for example, boys and girls are given a severe thrashing if
they are caught masturbating. In New Guinea, if a Kwoma woman sees
a boy with an erection, she will beat his penis with a stick. These
boys soon learn to refrain from touching their genitals even while
urinating.

These restrictive practices in primitive societies probably have a
familiar ring to western readers. This is because western society
(Europe and America) has, until very recently, also been restrictive
in regard to premarital sexual activity. By using similar procedures,
western societies have also attempted to prevent sexual intercourse
before marriage, prevent the young from acquiring full knowledge of
sexual relations, restrict girls' sexual behaviour much more than that
of boys, ensure virginity of the girls until marriage and also restrict
other forms of sexual behaviour such as masturbation and homosexuality.

It should be pointed out that these restrictive norms in primitive
tribes are not simply copied from western societies. Although influence from other cultures can never be completely discounted, the restrictive practices in these societies were by and large developed independently and are indigenous to their societies. The similarity
between the norms in these primitive societies and recent norms in
our modern society cannot be explained away as merely copies or reproductions due to contact with our culture. Many of the anthropological studies of these societies were carried out before western mis-
sionaries and traders were fully established. The explanation for
these similarities lies elsewhere.

However, most primitive societies are considerably more permissive concerning premarital sexual activity. Some tribes have formal
prohibitions against premarital affairs, which are, in fact, not taken
very seriously and not enforced. In such societies, children are able
to engage in sexual behaviour in secrecy. Even though the adults, of
course, know what is happening, they do not punish the children unless
it is brought flagrantly to their attention. For example, the Huichol
prohibit premarital sex, but adults do not keep close surveillance on
the young. The young people have ample opportunities to slip off into
the bush during dances and feasts. Among these societies adults sel-

dom are concerned with sexual experimentation of young people unless
pregnancy occurs. Among the Andamanese, should a girl become
pregnant, the parents arrange for the couple to be married.

A large number of societies are almost completely permissive towards sexual behaviour in children. In these societies children are
allowed or even encouraged to participate in sexual activities as soon
as they wish. In such societies babies are not discouraged from fin-
gering their own genitals. As children grow older, they are freely
allowed to handle the genitals of other children of either sex during
their play. The adults in the Pukapukans of Polynesia simply ignore
the sexual activities of children, boys and girls are allowed to mastur-
bate freely and openly in public. In a few such permissive societies,
adults actively participate in the sexual stimulation of infants and children. In the Alorese society, mothers sometimes fondle the genitals
of their babies while nursing them.

In these permissive societies, no attempt is made to prohibit the
children from acquiring sexual knowledge. Ponape children are given
careful instruction about sexual intercourse by the time they are four
or five years old. Sex knowledge is completely accessible to young
Alorese children and they are well informed of all the details by the
age of five. Often no special precautions are taken to prevent children
from witnessing adults sexual activities. Lesu children are allowed to
watch adults copulate, with the exception that they are not to observe
their own mothers having intercourse.

As soon as children in these societies are capable, they are allowed
to have full sexual intercourse. Early sex play usually involves many
forms of mutual masturbation, oral-genital contacts and usually ends
in attempted copulation. Most girls among the Lepcha of India engage
in full intercourse by the time they are eleven or twelve. Sexual intercourse begins among the Trobrianders at six to eight for girls, and
ten to twelve for boys. Among the Ila-speaking peoples of Africa,
there are no virgins after the age of ten.

So we see then, that there is a remarkable variation in the premarital sex norms in different societies. Why do these norms differ so?
It was proposed in the previous chapter that all norms in each society
are believed to be objectively functional by that society. How can such
a wide variety of norms all be accepted as functional by these different
societies? The next section will attempt to formulate an answer to
these questions.

SEX NORMS BASED ON HYPOTHESES
    It is now possible to consider how the sexual norms in all societies

(including our own) follow the principles which I proposed in the previous chapter. It was proposed there that all norms in all societies are
based on hypotheses which are accepted as valid by that society. A
great deal of the variability of norms in different societies can be attributed to the wide variety of magical, religious and scientific explanations accepted in these societies. It was also suggested that norms
also differ because of the differing environments from one society to
the next.

Norms are legitimized to encourage people to behave in a way which
it is believed (on the basis of a hypothesis) will have long-term survival
advantages for that society. Norms are necessary because of the conflict between the immediate consequences for the individual of an activity and the long term consequences of that same activity for society.
Through the use of auxiliary consequences, societies attempt to per-
suade people to act in a way which is functional for the future survival
of society.

Sexual norms in all societies are also instituted upon these same
principles. Sex norms are always based on hypotheses assumed to be
valid by that society. The wide variety of premarital sex norms in
different societies just discussed can be explained by the different
hypotheses accepted by each society and also by the differing environments of each society.

First of all, "marriage", that is a set of norms which encourages
two or more people to remain together for a considerable period of
time to provide for the care and protection of children, exists in all
societies. Marriage is universal because all cultures accept the hypothesis that human children need many years to mature. As mentioned
earlier, this extended "growing-up" hypothesis is actually only part of
the "birth-death" hypothesis discussed in Chapter 7. The idea that all
children will eventually become adults and then some day die is actually only a hypothesis.

Since marriage exists in all of today's primitive societies, it seems
likely that humans acquired this knowledge of our extended childhood
relatively early in our prehistory. The institution of marriage may
have existed in early human societies relatively soon after our ancestors acquired the capacity of "imagination" (the ability to hypothesize
and symbolize). Probably relatively soon after we were capable of
hypothesizing, early humans must have proposed that all human children needed a very long time to mature. Once this hypothesis had been
proposed and accepted, the legitimization of marriage norms was inevitable. Early human societies, like all of today's societies, must

have instituted marriage norms in an attempt to provide for the well-
being of their young in the future. Marriage is, and has been, universal in all societies which accept the hypothesis that children require a
very long period of care and protection by two or more adults.

After early humans had accepted this extended "growing-up" hypothesis, they must have then encouraged the notion that women should only
become pregnant and give birth within this institution of marriage.
However, there is good evidence for believing that for most of human
existence we were ignorant of the cause of pregnancy. As already discussed, several of today's primitive tribes have been found to be ignorant of this sex-pregnancy knowledge. It is possible that all humans
were ignorant of this sex-pregnancy relationship (at least in relation
to animals) until as recently as 10,000 years ago when we began using
domesticated animals as a major source of food. If humans were indeed ignorant of this relationship until only 10,000 years ago, this
would mean that for at least 99% of our existence, we were not aware
of the cause of conception.

Of course, this is not to say that early humans did not accept various magical or religious explanations as to why women became pregnant. All societies accept hypotheses concerning most events which
are considered important. Remember that both the Trobrianders and
the aborigines of Australia believed that women became pregnant because spirit-children entered the women. It seems most likely that
early human societies also accepted such religious or magical explanations.

Once humans had accepted the extended "growing-up" hypothesis,
they legitimized marriage norms; however, there is sound evidence
for believing that for many generations thereafter, humans, in fact,
remained ignorant of why women become pregnant. Early human societies must have discouraged women from having children outside of
marriage, but the problem was, they did not actually know why women
became pregnant.

All of today's primitive societies studied by anthropologists which
were ignorant of the sex-pregnancy relationship, were also extremely
permissive concerning premarital sex norms. It is my proposal that
these societies did not prohibit sex before marriage because they didn't
know that sex causes pregnancy. Societies prohibit activities which
they believe will have damaging consequences in the future. These
societies didn't prohibit premarital sex because they didn't know the
consequences of this activity. They saw no more harm in their children engaging in sexual activities than we do in our children playing

football. They didn't prohibit it because they didn't know the long-term consequences.

On the other hand, these early societies probably had other magical or religious hypotheses explaining why pregnancy occurs, they may well have enforced other restrictions on single girls' behaviour. For example, remember that the Trobrianders believed that girls became pregnant when spirit children entered them. It was believed that these spirit children were attached to drift logs, leaves or branches in the sea. Consequently, whenever much debris accumulated near the shore, single Trobriand girls would not enter the water because they were afraid they might become pregnant by doing so.

Therefore, if humans were ignorant of the sex-pregnancy relationship for most of our existence, we must also conclude that for most of our existence most human societies also did not prohibit premarital sex. Humans only prohibited premarital sex after we had learned the long-term consequences of this activity. Speaking in terms of our total existence on earth, premarital sex norms are most likely a very recent acquistion.

It is necessary to add some qualifications here. First of all, it is possible that a few societies might have prohibited premarital sex even though they were ignorant of the sex-pregnancy relationship. This would only be the case however, if they accepted a magical or religious explanation which predicted undesirable consequences for this activity. For example, if a tribe believed that sexual activity in children caused droughts or leprosy or such, they would, of course, prohibit this behaviour. Societies always prohibit activities which they believe, on the basis of a hypothesis, will lead to undesirable consequences.

A few such magical hypotheses have been recorded among today's primitive societies. For example, in the east central Carolines it is believed that intercourse before the first menstrual period may injure girls. Consequently, girls are strictly forbidden to have intercourse before puberty, but thereafter they enjoy almost complete sexual freedom. This hypothesis is not, as far as we know, valid. No harm is done to girls who copulate before puberty and of course it is virtually impossible for them to become pregnant.

Also, if a society was aware of the general sex-pregnancy connection, but had not yet worked out the precise nature of this relationship, this would mean that they would allow unusual activities from our point of view. For example, remember that the parents in Managaia (Marshall & Suggs, 1972, p. 130-1) encourage their daughters to "sleep around", that is going from man to man often until they find the man they wish to marry. This may seem like a rather unusual norm from our point of view, but it makes sense once you understand that the Mangaians accept the hypothesis that pregnancy only results from sleeping with one man regularly. Mangaian parents (like parents in all societies) do not wish their daughters to become pregnant before they are married, but because they accept an invalid hypothesis, they encourage a completely different behaviour in their daughters from that encouraged by parents who have a more precise understanding of the sex-pregnancy relationship.

The second qualification to my theory concerning sexual permissiveness is that although all of today's primitive societies which were ignorant of the sex-pregnancy relationship were also permissive concerning premarital sex the contrary is not the case. That is, many permissive societies were *not* ignorant of where babies come from. Even though some societies know why women become pregnant, they still do not prohibit premarital sex.

This can be explained, however, by the fact that a society's norms are not only based on the hypotheses which they accept, but also on the environmental situation of that society. Even if a society accepted the hypothesis that sex causes pregnancy, because of other factors in their society, the problem of illegitimate births might occur only rarely. For example, in societies which allow marriage at an early age, illegitimate births may very seldom occur. Because of adolescent sterility and early age of marriage, single girls might only rarely become pregnant even though they had a great deal of sex. In such societies, most of the girls would already be married before they were capable of conceiving.

Also, even if a permissive society accepts the sex-pregnancy relationship, they may accept other hypotheses (either magical or scientific) which propose that they are able to prevent pregnancy outside marriage. In several African tribes, adolescent boys are taught to practise interfemoral intercourse or coitus interruptus to avoid impregnating the girl. Some tribes believe that by orally ingesting certain medicines, they become temporarily sterile. Some societies may resort to abortions to deal with the occasional illegitimate pregnancy. In many permissive societies, the occasional illegitimate pregnancy is dealt with simply by requiring the couple to marry.

The conclusion then, is that all societies attempt to ensure that children will be born in an environment where their care and protection can be looked after for many years. For this reason, all societies have encouraged that children only be born within the institution of mar-

riage. As long as humans were ignorant that sex causes pregnancy,
they probably had no reason to prohibit sexual activities in young people.
Even after humans had accepted the sex-pregnancy relationship, they
probably did not outlaw premarital sex outright as long as the problem
of illegitimate births was relatively rare or could be dealt with in
other ways.

Premarital sexual activities are prohibited in the relatively few
restrictive primitive societies because they accept the two hypotheses
that (a) for their long-term welfare, children should only be born within
marriage, (b) sex causes pregnancy. In addition it seems likely that
experience has shown in these societies that other less drastic measures than prohibiting premarital sex were not effective in preventing
illegitimate births. Keeping young people from engaging in sexual
activities requires a great deal of time and energy on the part of adults.
It seems doubtful that a society would instigate this outright ban on sex
for young people unless other methods of preventing illegitimate births
had clearly been shown not to be effective.

It is probable that at least part of the reason why most of the primitive tribes were permissive and virtually all modern societies have,
until very recently, been restrictive, lies in the difference in the age
of marriage. Primitive tribes require much less training and education before their children can become self-sufficient. Young people are
able to master the skills of hunting or simple farming in a relatively
short time and consequently they become self-sufficient at an early age.
Thus, they are able to marry young. And as we have seen, very early
marriage goes a long way towards eliminating the problem of children
born out of wedlock.

In primitive societies where young people were allowed to marry
early, the problem of single girls getting pregnant probably occurred
only rarely, even though they were allowed as much premarital sex as
they wished. But as societies became more and more complex and
sophisticated, the age at which young people were able to support children gradually became higher and higher for a variety of reasons such
as the need for longer periods of training or education before they could
become self-sufficient. As the minimum age of marriage becomes
higher, so does the problem of illigitimate pregnancies. The old methods used in attempting to prevent the occasional illegitimate birth, such
as primitive forms of birth control, infanticide, abortion or forced
marriages were not sufficient by themselves to deal with the new surge
of single girls becoming pregnant. The more advanced societies have
had to prohibit premarital sex because the other methods were either

not acceptable on a large scale or else not fully effective.

Of course, there are other environmental factors involved in addition to the age of marriage. Fertility in different societies may vary due to many factors such as inbreeding, nutrition and racial characteristics. The period of adolescent sterility may also vary in extent and duration from one society to another for similar reasons. The extent of wealth or poverty, overpopulation or underpopulation must also affect a society's norms concerning premarital sex. For example, due to the terrible epidemics which ravaged Europe in the middle ages, the population in many places was reduced by half or even two thirds. It is estimated that it took nearly two centuries for Europe to regain the level of population which existed before the plagues began in 1348. During this subsequent period of severely reduced population, the sanctions against illegitimate pregnancies were also greatly reduced. People were much less troubled by illegitimacy. They did not actually encourage women to have children out of marriage, but on the other hand, it was much more readily tolerated. Obviously, the need for society to rebuild its seriously depleted population as quickly as possible was considered to be far more important than attempting to ensure the best possible care of some of these new members. As many children as possible were needed to rebuild the population, and society ceased to care, at least temporarily, whether they were born in or out of wedlock.

It is also possible that once a society had prohibited premarital sex (to prevent illegitimate pregnancies), some over-populated societies might have realized that artificially raising the age of marriage is one method of lowering population growth. Once a society had prohibited premarital sex and consequently effectively prevented most illegitimate pregnancies, the older a couple were when they married, the fewer children they were able to produce. Encouraging couples to marry late is one method of birth control. Some societies may have raised the socially acceptable age of marriage beyond the age actually necessary for the young people to become productively independent in order to limit population growth. Several societies including present day Ireland have used this method of birth control.

Modern societies have prohibited premarital sex because of the hypotheses they accept and the environmental conditions have meant that this prohibition was the best way of dealing with the problem of births outside marriage. The only other solution to this problem would be widespread use of abortion or infanticide for single mothers. This method has been used by a few societies, such as Iceland in the middle

ages, but abortion and infanticide are always used only as methods of
the last resort. Until recently, abortion was usually unsafe for the
mother, and infanticide is a traumatic and repugnant experience for
everyone involved. Most societies have decided that prohibiting premarital sex was a far better method of dealing with the problem of illegitimate pregnancies than the use of widespread abortion or infanticide.

Modern societies have prohibited premarital sex for the same reasons as the restrictive primitive societies. The similarities that exist
between our norms and theirs such as attempting to prevent sexual
intercourse before marriage, preventing the young from acquiring full
knowledge of sexual relations, restricting girls' sexual behaviour much
more than that of boys, ensuring virginity of the girls until marriage
and also restricting other forms of sexual behaviour such as masturbation and homosexuality, occurred because of the similarities between
the hypotheses accepted and environmental conditions. It was not
chance or cultural borrowing which produced these similarities, it was
a similarity of causal factors.

Viewed in this way, the reason why many societies restrict girls'
premarital sexual behaviour more severely than that of boys becomes
clear. If the function of the norms is to prevent children being born
out of marriage, then it is evident that girls should bear the brunt of
the restrictions. The girls are the ones that become pregnant. Although there may be some doubt who the father is, there is never any
doubt who the mother is. And it is the mother and her family who
must deal with the problem of bringing up the illegitimate child.

The "double standard" towards premarital sex norms exists in many
different restrictive societies because (a) females are the sex that get
pregnant and they must face the problem of bringing up the child, (b)
perhaps many societies reckon that if you could concentrate on preventing single girls from copulating, then of course, the boys will have no
one to copulate with, (c) it is easy to identify and punish the mother of
an illegitimate child, but this is sometimes not so with the father. No
doubt if boys were the ones that became pregnant, the double standard
would have been reversed.

Restrictive societies attempt to ensure that girls are virgins before
they marry because a virgin bride, by definition, has never had premarital sexual intercourse and this of course means that it is impossible for such girls to become pregnant before they are married.
Virginity at marriage is a functional norm because it ensures that there
are no illegitimate births.

LEGITIMIZATION OF SEX NORMS

If virtually all early human societies were permissive concerning premarital sex because they were ignorant of the sex-pregnancy relationship, then it follows that once early man had proposed and accepted the sex-pregnacy hypothesis, this might sooner or later necessitate a change in his customs, taboos, morals or laws. Like other rules, sexual norms are dependent upon the hypothesis accepted and the environmental conditions found in each society. Once an early human tribe had changed their hypothesis concerning pregnancy, this might well have eventually required a change in their norms.

Since many of today's primitive societies are aware of the sex-pregnancy connection, but nevertheless are still permissive concerning premarital sex because illegitimate pregnancies occur only rarely, then the early human tribes which first discovered where babies came from may also have found it unnecessary to change their sex norms immediately. However, as this knowledge spread to more and more tribes living in different environments and as societies gradually became more and more technically advanced, it would sooner or later occur that a society which accepted the sex-pregnancy connection would also have problems concerned with a large number of illegitimate pregnancies. Since all societies encourage the notion that children should be born within marriage, and since this tribe also accepted the sex-pregnancy relationship (or more precisely that they accepted that any one act of intercourse could result in pregnancy) there was only one possible solution to this problem. At this point the elders of the tribe, after much talk and discussion, must have decided that it would be a good thing if young unmarried people did not engage in sexual intercourse. In order to prevent widespread illegitimate births or else widespread abortions or infanticide, it was decided that intercourse outside of marriage was a "bad" thing (i.e. it was not advantageous to the long-term survival of that society). No doubt this discussion was couched in terms of the existing magical or religious belief system, however.

Once they accepted that premarital sexual intercourse was harmful to the survival of their tribe, they may have patiently explained to the young people why this was the case and asked them to refrain from this activity until they were married. The single people most likely accepted the reasoning of their elders, but as we saw in the last chapter, people sometimes have a great deal of difficulty changing their behaviour in the desired direction even when they accept the long-term advantages of doing so.

To the individual, the immediate consequences of an activity often outweigh the remote long-term consequences of the same activity. This is particularly true in the case of sexual intercourse. Sexual intercourse is probably the most enjoyable single activity known to humans. The immediate consequence of intercourse is extreme pleasure, but the long-term consequences, at least for single people, are undesirable for the society. This sets up a conflict situation for the single people involved, and as proposed in the previous chapter, this conflict between the immediate or personal consequences versus the long-term consequences for society is the basis for all norms or rules.

Even after the single people had accepted the logic of avoiding intercourse before marriage, they must have had a great deal of trouble changing their behaviour. To the young person, the immediate advantages must have easily overpowered the long-term disadvantages. At this point, the elders of the tribe must have realized that simply informing the young people was not sufficient, that many of the single people were unable to alter their behaviour even after they accepted the reasons for doing so. The high incidence of illegitimate pregnancies continued.

The elders must have then reluctantly decided that a norm or rule would have to be legitimized, that is auxiliary consequences and values added to help the single people change their behaviour in the desired direction. New immediate and undesirable consequences would have to be added to counteract the immediate desirable consequences of sexual intercourse. In short, it was necessary to discourage premarital sexual activity by the use of a variety of punishments and positive and negative cultural values.

Since sexual intercourse produces immediate and extreme pleasure in the individual (evolution surely made it so to ensure reproduction of the species), these early societies which attempted to prevent intercourse in single people must have soon found that in order to successfully reduce the incidence of premarital sex, it was necessary to legitimize strong negative auxiliary consequences. As discussed in the previous chapter, societies must fight fire with (an opposite but equal) fire. The powerful immediate desirable consequences had to be overcome by equally powerful undesirable consequences.

As those early societies (as well as all subsequent restrictive societies) must have discovered, it is not easy to suppress the instinctive sexual urge in young people. Strong social, moral and sometimes even legal sanctions proved necessary. The religious leaders must have declared that this behaviour was strongly disapproved of by the

relevant gods and that punishment would occur in the person's proposed
afterlife. Young people who continued to "misbehave" were severely
chastised, disgraced or even rejected by their family and tribe. In
some cases legal sanctions may have been used against offenders.
Physical punishments may have been used. In some of today's primitive tribes, if a couple are caught copulating they are beaten. Also,
remember that in the Gilberts, if knowledge of premarital sex becomes
public, both people are put to death. Many examples of similar social,
moral and legal punishments can be found in the literature and history
of western societies.

It was also necessary to assign strong positive cultural values to
virginity at marriage (especially for girls) and equally strong negative
values equivalent to the English words "sinful", "evil", "dirty" and so
on to premarital sexual activities. Since most societies deemed it
more important to ensure the girl's virginity at marriage (as opposed
to the virginity of boys), different cultural labels were assigned to the
different sexes for the very same behaviour. Boys who engaged in premarital sex activities were only mildly reprimanded with phrases such
as "boys will be boys" or "sowing their wild oats". However girls who
engaged in premarital sexual activities were assigned strong negative
labels such as "slut" or "whore".

The strength of these negative auxiliary consequences and values
probably also varied from culture to culture depending upon a number
of factors such as (1) the number of illegitimate births occurring in
that society (and this depends on many environmental factors including
the usual age of marriage) and (2) how difficult it would be for a single
girl and her family to raise a child on their own in that society. Thus
a society in which say 50% of the single girls were becoming pregnant
would probably legitimize much stronger negative auxiliary consequences than a society in which only 10% of the girls were becoming
pregnant. And a society in which, due to environmental factors, it
would be extremely difficult for a single girl to raise a child would
probably enact stronger auxiliary consequences than a society in which
it was not so difficult to do so. Thus restrictive societies may differ
widely in the strength and severity of the auxiliary consequences due to
many environmental factors.

These early societies may well have used a variety of other strategies in attempting to prevent young people from engaging in premarital sexual intercourse. It is common in many of today's restrictive
primitive societies (as well as in modern societies) to enforce segregation of the sexes in young single people, strict chaperonage (espe-

cially of the girls), attempting to keep the young people in complete
ignorance of sexual relations (the tactic here apparently being that if
young people didn't know about sex, they wouldn't engage in it) and if
all these methods fail forcing a couple (assuming the father was known)
to marry if the girl becomes pregnant. These very same methods may
well have been used by the earliest societies in attempting to eliminate
illegitimate births.

Of course, we do not really know what methods these early societies
used in attempting to restrict premarital sexual intercourse, the above
description is only a theoretical reconstruction of what must have happened, but we do know that all of today's restrictive societies (primitive and modern) use some or all of the above punishments, values and
strategies in attempting to prevent intercourse before marriage. It is
doubtful if any of these methods were or are 100% successful in enforcing chastity in young people. No matter how severe the auxiliary
consequences, a few "mishaps" always occur. However, these auxiliary consequences are effective in *reducing the incidence* of the undesirable behaviour. All norms (sexual or otherwise) seldom if ever
achieve total compliance by the people, but they are nevertheless often
successful in severely reducing the rate of "misbehaviour". These
norms continue to be enforced because a low rate of incidence is better
than a high rate. No doubt a society would prefer to have no illegitimate pregnancies, but failing that, a small number of pregnancies out
of wedlock is better than a large number. Norms do not eliminate
undesirable consequences, but they keep them down to a manageable
size.

No doubt the functional reason for legitimizing a norm against premarital sexual intercourse was clearly evident to the elders of the
tribe who first instituted this restrictive practice. The need to reduce
illegitimate pregnancies was evident, and armed with the knowledge of
the sex-pregnancy connection, the remedy seemed obvious. After this
norm had been legitimized, however, the functional nature of this rule
began to fade. As discussed in the previous chapter, after a norm has
been legitimized, the society often begins to forget or in some cases
even suppresses the functional nature of the norm. A law becomes
accepted as valid simply because "it is the law" and a moral simply
because "it is wrong". As today's sexually restrictive societies usually
rely heavily on the religious "moral" type of norm to control sexual
behaviour, the first premarital sexual prohibitions may also have been
a moral type of norm and consequently the functional nature may well
have been actively suppressed in this early society. For reasons

already discussed, society must actively suppress the functional nature
of morals or taboos to help ensure their compliance by the people.

Although the legitimization of a rule against premarital sex and the
subsequent forgetting of the functional nature of this norm must have
first occurred in some nameless pre-historic society, these same
causal factors explain the existence of restrictive sexual norms for
single people found, not only in some of today's primitive societies,
but also in modern societies. Modern societies have attempted to prohibit sexual relations in young people for the very same reasons other
societies have. Until very recently all modern societies have attempted
to restrict this behaviour for as far back as our records go because
(1) we accept the sex-pregnancy hypothesis, and (2) our ancestors
found that without this norm there was a large incidence of illegitimate
pregnancies.

And like the first restrictive society and today's restrictive primitive societies, we have forgotten or suppressed the functional nature
for instituting this norm. To ensure compliance with this moral, it
was necessary for the people to accept the idea that this behaviour was
wrong because "the Bible says so" or because "it is wrong" or such.
To be effective, the people must believe that religious morals are
backed up by supernatural rewards and punishments. If the functional
nature of a moral was admitted, it would open the door to doubting the
existence of these supernatural consequences. We are ignorant of the
real functional nature of our sexual norms for the same reasons the
Hindus of India are ignorant of the functional reason for their norm
against killing cows. In both cases the functional nature of the norm
has been suppressed to ensure compliance, leaving only the religious
justifications of "cows are sacred" or "it is a sin".

These restrictive sexual norms have existed therefore not because
of *religion*, but rather because of *science*. Many young people in western societies today decry the Christian religion for instituting these
restrictive sexual morals for many centuries. But many other religions have instituted very similar morals. Even modern secular com-
munist societies such as the Soviet Union and China have discouraged
this behaviour by the use of other types of norms. China is today one
of the most sexually puritanical societies in the world. The root cause
of these norms was not religion, but rather science. Religion has only
been the justification for these rules. The actual cause of this norm
was a pre-historic scientific discovery-- namely that sex causes pregnancy. Religion only provided the auxiliary consequences to enforce
compliance of this rule. If young people want to blame an institution

for the existence of these restrictive norms they should blame science
rather than religion. Religion only came in afterwards to enforce the
rules which science said were necessary.

Although these restrictive sexual norms may seem harsh and unnecessary to young people in western societies today, they were deemed
to be the best solution to a very serious problem which existed in these
earlier societies. To a less advanced society, the problems created
by many single girls giving birth without a man to help provide food,
shelter, clothing and so on for many years, were enormous. In a primitive society, a single girl might well have extreme difficulties attempting to bring up a child without a husband. This would be true especially
if her parents and relations were no longer living or too old or unable
or unwilling to help with the child. To a society living close to the
subsistence level, an illegitimate birth may well have put an unbearable
burden on the girl and her family. And short of widespread abortions
or infanticide, the only effective method of dealing with this problem
for a tribe which knew the sex-pregnancy connection was to prohibit
sexual activities in young people. Thanks to a greatly improved standard of living in modern societies, the problems posed by illegitimate
pregnancies are not nearly so great today. But today's young people
should remind themselves that most of this improvement has come
about only in this century. Harsh though they may seem to us today,
these restrictive sexual norms were believed to be objectively functional and necessary in these earlier societies. Due to scientific advances these restrictive norms are changing in today's societies, but
this matter will be discussed in detail in the next section.

There is one lingering question which some readers may be asking
themselves. Granted, prohibiting premarital heterosexual intercourse
was functional and necessary to prevent illegitimate pregnancies, but
why is it necessary to prohibit other forms of sexual activity such as
homosexuality and masturbation as well? Many restrictive societies,
including our own, have prohibited these activities. These activities
do not result in pregnancy, so why do some societies restrict them?

Although many permissive societies believe that masturbation in
young children is a normal activity, as the child grows older, it is
assumed that this self-stimulation should be replaced by sexual activities with other children. Masturbation in adults is often regarded as
something to be ashamed of, done only by someone who is unable to
obtain a lover. Homosexuality is often freely allowed in most permissive societies, however in some permissive societies, definite social
pressure is directed against such behaviour (Ford & Beach 1951, p. 129).

The reason most restrictive societies prohibit these activities is that after a society had strongly prohibited premarital heterosexual intercourse, the young people may well have increased their activity of alternative forms of sexual satisfaction. Out of frustration, the incidence of other forms of sexual activity in single people may well have increased sharply. If a society severely restricts heterosexual activity, the young people might have well attempted to satisfy their sexual instinct by greatly increasing their activity of homosexuality and masturbation. It is possible that the elders may have worried that if young people became too accustomed to these alternative forms of behaviour before they married, they might prefer these to heterosexual activity when they came of age to marry. A society in which none of the young people engaged in heterosexual activity, they reasoned, would soon die out.

In a restrictive society which had accepted that sex outside of marriage was wrong simply because "the god(s) say so", it would be easy for them to include other forms of sexual behaviour in this prohibition. Alarmed by the sudden increase in these other forms of sexual gratification, and no longer clearly aware of why premarital heterosexual activities had been prohibited in the first place, it was easy for the elders to expand the ban on premarital heterosexual activity to *all* sexual activity outside of marriage.

Exactly why some permissive societies also discourage homosexual activities is not clear. Even some permissive societies which are ignorant of the sex-pregnancy relationship attempt to restrict homosexual behaviour. The people of these societies justify these restrictive norms against homosexuality with such reasons as "this man copulates excrement" and that homosexuality is an inadequate and contemptible substitute for heterosexual behaviour. Although as we have seen, the people in a society are not always aware of the actual functional reason for the existence of their norms. Thus these explanations may or may not tell us something about actual reasons for the existence of these restrictive norms.

It is also quite possible that the existence of "prostitution", sexual intercourse traded for a price, only occurred on a large scale after humans had prohibited sexual activities outside of marriage. Before the legitimization of these restrictive sexual norms, there was probably little if any need for "prostitution". In a permissive society, the young people could freely engage in sexual activities with each other, and if married people wished to engage in any "extracurricular" sexual activity, they could probably find ample opportunities for adultery.

But when prohibited from engaging in any form of sexual behaviour, single people are likely to be very sexually frustrated. The safety valve which prevents these frustrations from becoming intolerable (at least for men) is the institution of prostitution. Monetary rewards are given in compensation to the girls who break the strong moral sanctions against intercourse outside of marriage. The reason prostitution is usually "one sided" (that is, women catering to men, and only rarely men catering to women) is because it is the women who become pregnant. Women prostitutes accept payments not only to break the social norms, but also to accept the risks of pregnancy. Men prostitutes seldom exist because any women clients would not only have to pay the man for sexual satisfaction, but the women would still be stuck with the social disapproval if she became pregnant. A woman who could not contain her frustration could probably obtain sexual satisfaction without having to pay a man. Because the social norms affected the women much more strongly, however, many women were forced to deal with their frustrations in some less direct way.

Prostitution probably exists in all restrictive societies because a total ban on all sexual activities outside of marriage is almost impossive to maintain. What often has been described as "the world's oldest profession", is in all likelihood, a relatively *new* profession. At least in terms of our total existence on earth, the existence of prostitution is probably a recent occupation. Only after humans had prohibited sex outside marriage, did the existence of prostitution become necessary at least on a large scale. As already discussed, these norms could not have existed before humans learned the sex-pregnancy relationship and this may have occurred as little as 10,000 years ago. In all likelihood, the occupation of "hunter" is just as old, if not a much older occupation than "prostitute".

In conclusion, it seems likely that the legitimization of these premarital sex norms had an important effect on the status of women in these societies. For reasons already discussed, most restrictive societies decided it was more important to guard the virginity of girls before marriage rather than that of boys. These additional restrictions on the behaviour of girls such as strict chaperonage and much more severe punishments for premarital sexual activities probably reduced the status and even self-confidence of females in these societies. Men and boys continued to have the freedom to go where they wanted when they wanted, but in these restrictive societies females now were greatly curtailed in their freedom of movement and variety of activities.

Probably because of their greater physical size and strength, men

are the dominant sex in all known societies. However, in some primitive societies women are treated much more as equals and indeed within
their work areas or specialization (usually concerned with the home
and children) they are considered independent and even dominant.
There are some exceptions, but generally speaking, the sexually permissive societies appear to be much more egalitarian in the status of
the sexes than do the restrictive societies. Restricting the sexual behaviour of girls more than that of boys may well have had secondary
consequences concerning the status and position of females.

Also, until very recently, married women have been unable to control their fertility effectively. In a primitive society, pregnancy probably did not greatly interfere with a woman's ability to do her work and
even become an expert in any area she cared to pursue (except areas
such as hunting or tasks requiring great physical strength), but as societies became larger, more complex and advanced, more and more
apprenticeship or education and work away from home were required by
any young person wishing to enter a skilled profession. The fact that
women could not effectively control their own fertility meant that they
were often excluded from these professions. This problem must have
also added to the inferior status often given to women in the more advanced societies. Both these problems are now being overcome in
modern societies due to the very recent introduction of effective birth
control. This matter will be discussed in more detail in the next section.

THE "SEXUAL REVOLUTION"

Very recently in modern industrialized societies there has been a
marked change in the norms governing premarital sexual behaviour.
Modern societies, particularly since the 1960s, have become much
more liberal in allowing young people to engage in sexual activities
outside marriage. This rapid change in norms has often been referred
to as the "sexual revolution". This section attempts to explain why and
how these changes have occurred. I will attempt to show that these
changes are the result of the same principles which govern the formation and change in other types of norms, that these changes in norms
are a direct result of other earlier changes which occurred in these
societies. This section will propose that the "sexual revolution" is not
a sign of decadence, decay and moral laxity in modern societies, but
rather is a natural, lawful and even predictable change which occurs as
a result of changes in a society's hypotheses or environmental conditions.

It has already been proposed that premarital sexual rules are instituted only *after* a society has (1) accepted the sex-pregnancy hypothesis (in particular that any one act of intercourse could result in pregnancy) and (2) due to environmental factors large numbers of girls become pregnant before they marry. Once a society has acquired both of these characteristics, the legitimization of restricitive premarital sex norms becomes inevitable. In short, once a society has a problem (in this case a large number of illegitimate pregnancies) and once a society thinks it knows how to prevent this problem (in this case the sex-pregnancy hypothesis) it will always attempt to do so. In this section I will propose that these norms will eventually be liberalized or abandoned completely in any society which ceases to experience either of these two causative factors. Once a society changes its hypothesis concerning pregnancy or once a society ceases to have problems with illegitimate pregnancies, these restrictive norms will eventually and inevitably be changed.

Ever since early humans first accepted that there was some sort of connection between sex and pregnancy they probably also attempted to find some method of preventing this necessary connection. Evidence indicates that most if not all societies have attempted to find some way of invalidating the sex-pregnancy relationship at least part of the time. They have tried to find some way of having sex which does not produce children, both as a method of preventing illegitimate pregnancies before marriage and also as a way of limiting population growth after marriage. In short, they have attempted to find some method of what we now refer to as "birth control".

Before humans accepted the sex-pregnancy hypothesis, the effective methods of birth control open to them were very limited. As discussed in Chapter 7, since early humans did not yet actually know what caused the child to start growing inside the mother, they could not prevent this process from occurring. The only really effective methods of birth control available to early humans were infanticide and abortion. It is also conceivable that some early tribes might have learned that prolonging lactation sometimes provides a period of infertility. Even if some early tribes did accept this method, we know now that some women do ovulate during prolonged lactation, hence this method is not reliable. It has also been reported that some Australian tribes performed ovariotomies on some girls to prevent pregnancies (see Himes, 1963, p. 40-1). It is conceivable that some early tribes could have used this method of control without knowing the sex-pregnancy connection. All peoples, of course, know that the child grows inside the

mother, and it is possible that a tribe could have simply learned that
after an ovariotomy, a woman never became pregnant. It would not
have been necessary for them to know the man's role in conception in
order to use this method. Even if an early tribe did learn this method,
it would have had serious drawbacks due to lack of anesthesia, risks of
infection and that this is a permanent rather than temporary method of
birth control.

Of course, early humans may well have attempted numerous other
methods of birth control before they learned the sex-pregnancy relationship, probably these were related to the explanations they accepted
as to why women become pregnant. Remember that Trobriander girls
who did not want to become pregnant, refused to enter the sea when
drift logs, leaves or branches accumulated near the shore because they
believed "spirit children" were attached to them. Needless to say,
since such methods were based on invalid, magical or religious explanations, it is extremely likely that these methods were completely useless as methods of birth control.

However, once humans had accepted the sex-pregnancy connection,
a myriad of new possibilities for preventing pregnancies must have
occurred to them. Douching, coitus interrruptus, coitus reservatus,
temporary abstinence during part of the woman's period, suppositories
and barriers placed in the vagina before intercourse, castration, potions taken by either men or women, celibacy, perhaps even primitive
forms of condoms and vaginal caps as well as many other magical or
invalid methods may all have been tried by early humans in an attempt
to prevent unwanted pregnancies (both before and after marriage).
Certainly today's primitive tribes studied by anthropologists have attempted to use many of these methods. An interesting history of birth
control methods can be found in Himes (1963) or Draper (1965).

For example, several primitive tribes have attempted to prevent
pregnancies by placing a barrier in the vagina before intercourse. The
Dahomey tribe of West Africa used crushed tubercled root, in Easter
Island they used seaweed and in central Africa rags and grass. One of
the most effective barriers was used in Egypt at least three thousand
years ago. This consisted of crocodile dung. The dung provided a plug
and the acid in it acted as a spermicide. However, one disadvantage of
this method occurs if the plug cannot be removed, when the results
can be fatal.

Coitus interruptus (withdrawal before ejaculation) was almost certainly used by early humans once they learned the sex-pregnancy relationship. Evidence suggests that it was used in many parts of the

world including parts of Africa, the South Sea Islands and in Persia.
It is quite possible that this method achieved more fertility control
through the ages than any other single method. This method is only
partially effective however, but except with very fertile couples, it
generally succeeds in at least spacing the births.

It was only after humans had accepted the sex-pregnancy relationship, that it became apparent that celibacy was a method of preventing
pregnancies. Indeed, celibacy is a completely safe and 100% effective
method of preventing conception. The only trouble is most people, and
especially young people, find voluntary abstinence from sexual relations
for long periods very frustrating and difficult to maintain.

Since celibacy is so difficult to maintain, many peoples have tried
to find and use some alternative form of sexual gratification which
would not result in conception. Oral and anal intercourse, interfemoral
intercourse (in which the penis is inserted between the thighs rather
than in the vagina), homosexuality and special postures have all been
resorted to in different societies. The disadvantage of these methods
however is that they require a very high degree of self-control by the
participating individuals. When sexually aroused, the temptation to
resort to normal intercourse is strong indeed.

Early humans may also have hypothesized that there were certain
times during a woman's menstrual cycle when she was infertile, and
that by temporarily abstaining during this "safe period", a couple could
prevent pregnancies. This method has been attempted by the Nandi
tribe in East Africa and the Isleta Indians of New Mexico. The Greek
physician Soranus in the second century A.D. described this "rhythm
method" of birth control, but we now know that his recommended safe
period was only partially safe. This knowledge was apparently lost
during the middle ages and had to be rediscovered later. Knowledge of
the exact dates necessary to avoid intercourse were confirmed only in
this century. Previously, most doctors in Europe believed that the
rate of conception was highest immediately after menstruation. Even
today this method of birth control is unreliable. For women with
regular cycles this method is only partially effective, but for women
with irregular periods it is useless.

Also the possibility of castration as a method of birth control could
have been realized once a society accepted the man's role in starting
pregnancy. There is evidence that castration was used in ancient China
as a method of birth control. However, testes produce not only sperm
but also male hormones. Humans castrated before puberty never go
through puberty and are never capable of ejaculation. Castration in

adult men, often, but not always, results in a gradual reduction in the
desire and capacity for intercourse.

Douching by women after intercourse may also have been used by
early tribes. In South America lemon juice was mixed with the husks
of mahogany nuts and used as a douche. In both Greece and Rome
vinegar was used. Both lemon juice and vinegar are now recognized as
effective spermicides. This knowledge was also lost during the middle
ages and only reintroduced in the nineteenth century by an American
doctor, Charles Knowlton. Douching, even with modern products, is
not reliable, however.

It is even possible that early peoples may have tried primitive forms
of vaginal caps or condoms. Apparently oiled disks of bamboo tissue
paper were used as caps in ancient Japan and later, disks made from
melted bees-wax were used by German-Hungarian women. A female
condom was used by the Djukas tribe in South America, a seed-pod with
the end cut off was placed in the vagina before intercourse. The Romans may well have used some form of male or female condom made
from a goat's bladder. Dr. Fallopius apparently rediscovered the
condom (a linen bag) in the sixteenth century. It was gradually improved
by using various types of animal gut. It was only in the 1880's with
the vulcanization of rubber that rubber condoms were made.

Many primitive tribes have attempted to find some potion which
ensures temporary sterility. Numerous concoctions have been recorded both from primitive tribes and from our own history. At one
time or another people have tried consuming foam from a camel's
mouth, various types of roots, walnut leaves, yolk of eggs, gunpowder,
quicksilver, oil, saffron and so on. Even today many women in Colum-
bia believe that a mixture of lemonade and aspirin is effective in preventing pregnancy. Some of these invalid or magical potions were
based on what Frazer called the principle of similarity (discussed in
Chapter 8). Examples of these include drinking water which had been
used to wash a dead person and the eating of bread with honeycomb
containing dead bees.

Most of these potions must have been completely ineffective in preventing conception; however, it is possible that some of these early
attempts may have at least partially lowered women's fertility. For
example, one ancient potion was to drink the water in which black-
smiths had cooled their forceps. The water would have contained a
high level of lead and lead apparently has sterilizing properties (but
lead is also a poison).

It was only in 1955 that humans at last found an effective and rela-

tively safe oral contraceptive. Dr. Gregory Pincus found a group of
chemical steroids which inhibited ovulation. The earlier attempts at
using roots, herbs and leaves may not seem so far-fetched when we
learn that the basic ingredient of today's contraceptive pill is obtained
from the Barbasco root found only in Mexico. The contraceptive pill
was first put on the market in 1961.

Early humans, both before and after they learned the sex-pregnancy
relationship, probably also tried many other invalid or magical attempts at birth control which have no modern counterpart. For example, Maori medicine men would throw blood from the afterbirth on
a fire in the belief that this would prevent more pregnancies. Many
peoples wore various types of amulets containing such things as the
tooth of a child, part of a cat's liver, the testicle of a weasel and so on.

The important point here is that even after humans had learned the
sex-pregnancy relationship, most of their early attempts at birth control were either unreliable, unsafe or unacceptable to most people. It
is only in this century that we have perfected several of these methods
to the point where they are acceptable, efficient and safe.

Before this, some methods such as coitus interruptus, the rhythm
method, douching may have been partially effective in some societies
in limiting population growth and spreading births within marriage, but
in order to prevent as many illegitimate pregnancies as possible, a
more effective method was needed. As already discussed, abortion
and infanticide are usually used only as methods of the last resort to
limit further population growth, and the widespread use of either of
these methods to eliminate illegitimate births was unacceptable to most
societies. Ovariotomies and castration (or more recently vasectomies)
were unacceptable means of birth control for unmarried people because
they were permanent. If all young people were sterilized to prevent
illegitimate pregnancies, they would continue to be sterile after they
married and thus the society would die out. No, the only effective and
safe method of preventing illegitimate pregnancies once a society had
learned the sex-pregnancy connection was celibacy for single people.
This was the only completely safe, effective and acceptable method of
birth control known until very recently.

Thus, the elders of the first prehistoric society to prohibit premarital sexual activities must have reluctantly decided to ban this behaviour because all the other methods of birth control known to them
had been found unsuitable for one reason or another. Prohibiting premarital sexual activity is equivalent to advocating celibacy for single
people. Early tribes may well have tried other methods of birth con-

trol before restricting sexual activities in single people, but found all
of them unsuitable.

Enforcing celibacy on young people requires a great deal of time and
energy by the adults, and it seems likely therefore, that they would have
resorted to this method only after they had tried and rejected all the
methods less difficult to enforce. Indeed, the only major drawback to
this method of birth control is that the young people find enforced celibacy very frustrating. If this society had had another effective, safe
and acceptable method which was less frustrating and less difficult to
enforce, no doubt they would have used it.

But once the norm had been legitimized, the functional reason why
celibacy in single people was a "good" thing was suppressed. Celibacy
and virginity became "good", "holy" and "pure" for no apparent functional reason. Likewise, sexual activities in single people became
"dirty", "sinful" and "wrong" simply because the god(s) declared it to
be so. The fact that celibacy was the only suitable method of birth
control these early societies had for single people was simply ignored
and eventually forgotten.

It may well be that once the people had accepted that celibacy was a
"good" thing for its own sake in single people, this positive ideal of
celibacy eventually spread into other areas of life. In much of Christianity, as well as some other religions, the person who remained celibate throughout his life is thought to be holier or nearer to god than
other people. St Paul held that sexual intercourse, even within marriage, might prove to be a handicap in the attempt to win salvation
(I Cor. VII, 32-34). Once the functional reason for celibacy in single
people was lost, the ideal of celibacy as "good", "holy" and "pure"
may well have generalized to other situations.

Also, since other methods of birth control had already been tried
and found lacking, societies may well have attempted to keep the knowledge of these other forms from the young people. Not only did many
restrictive societies not tell their children anything about sex, but they
tried to keep other forms of birth control secret as well. If a young
person knew about other methods of attempting to prevent conception,
it would only encourage him to try intercourse using these methods,
and these methods had already been found unsuitable for one reason or
another. Because all the other methods had at least one flaw, the use
and even knowledge of them should be suppressed and made "immoral"
for single people. And again, after this norm was instituted, the functional reason was forgotten or suppressed. Other forms of birth control became "wrong" for single people simply because the religious

leaders said so. Among at least some groups of Christians this norm against using other forms of birth control may also have generalized to within marriage as well as outside marriage. Certainly the Christian church strenuously opposed most of these methods of birth control both in and out of marriage until recently. However, the more practical advantage of producing as many new people and believers as possible may also have been responsible for this ideal of not using most other forms of birth control within marriage.

This situation in western societies continued along with some variation from one society to another (due to numerous factors) until very recently. Virtually all western societies have discouraged sexual relations in single people with greater or lesser severity for as far back as we have records. However, beginning as early as a century ago, science began to make some advances in the field of birth control, and these improvements were eventually to alter our sex norms radically.

As already briefly reviewed, science recently began to make marked improvements in the technology of birth control. Although religious bodies often opposed these improved techniques and tried to prevent this new knowledge from becoming widespread (because religious people were no longer aware of the reasons why other forms of birth control had been pronouned "wrong"), gradually this improved knowledge and equipment became widely disseminated. Improved condoms, caps and spermicides as well as more exact knowledge of a woman's "safe period", meant that early in this century by careful use of one or more of these methods, couples could effectively plan when and if they had children. The culmination of this improvement occurred in 1961 when the contraceptive pill was introduced. Except for celibacy, this was the first 100% effective, relatively safe, acceptable and temporary method of birth control available to humans. After a pre-historic scientific discovery, humans had realized that celibacy was an effective means of birth control, and after another, modern scientific discovery, humans had finally found another suitable method.

These very recent improvements in other methods of birth control were eventually to change our sexual norms. Once there existed other suitable methods of birth control, it was no longer necessary for modern societies to prohibit sexual activities before marriage. Another suitable, but less frustrating and less difficult to enforce, method had at last been found. Since the functional reason for instituting premarital sex prohibitions had been to reduce illegitimate pregnancies, this same function could now be served by other methods. The function remained the same, only the methods changed. The sexual revolution has only meant that single people in modern societies have switched from using one method of birth control to using another.

In effect, these improved methods of birth control meant that modern societies had changed their hypotheses concerning pregnancy. Previously, our society had accepted the hypothesis that no matter what preventive measures might be attempted, any one act of sexual intercourse could result in pregnancy. In short, sex sometimes results in pregnancy. Now, however, with modern methods of birth control, we accept the hypothesis that with proper preventive measures, sex will not result in pregnancy. In short, sex and pregnancy are no longer invariably related.

In the first part of this century, the changes in our sex norms resulting from these improvements were almost imperceptible. However, these changes were slowly occurring and at least one observer even predicted our "sexual revolution". Terman (1938) interviewed married couples concerning, among other things, premarital sex behaviour. He found considerable differences between the older and younger couples in this matter. He noted that the proportion of men and women who were virgins at marriage had steadily decreased between 1910 and 1930. Terman remarked that if this average rate of decrease continued into the future, this would mean that "virginity at marriage will be close to the vanishing point for males born after 1930 and for females born after 1940. It is more likely that the rate of change will become somewhat retarded as the zero point is approached and that an occasional virgin will come to the marriage bed for a few decades beyond the dates indicated by the curves. It will be of no small interest to see how long the cultural ideal of virgin marriage will survive as a moral code after its observance has passed into history" (p. 323). Terman's prediction turned out to be remarkably accurate at least in Europe and America.

The final death blow for our restrictive sexual rules occurred in 1961. The contraceptive pill was the first 100% effective, relatively safe (and this safety has been improved since then) and temporary method of birth control discovered since pre-historic humans had discovered that celibacy was a method of birth control. It is no accident therefore that the sexual revolution became much more pronounced in the 1960s. The sexual revolution was an inevitable result of these scientific improvements in the field of birth control.

The norms in a society are invariably linked to the hypotheses accepted by that society. This is no less the case with our sex norms than with other norms in distant societies. Early human societies did not prohibit premarital sex because they had not yet discovered the sex-

pregnancy connection, that is, they did not realize the long-term consequences of this activity. Modern societies are now beginning to
condone premarital sex again because we have at last learned other
ways of preventing these consequences. Thus, both the instituting of
these restrictive norms in some pre-historic society and the abandoning
of these same restrictive rules in modern societies are the result of
scientific discoveries.

As with other changes in social norms, there existed a considerable
"cultural lag" in this area. Even after discoveries and improvements
in the field of birth control had been made, some people attempted to
prevent these advances from becoming widely known and distributed,
and even after this occurred, they continued to label premarital sex
behaviour as "wrong", "immoral" and so on. One reason for the cultural lag in this area is the result of the phenomenon of the forgotten
function discussed in the previous chapter. Since most of the people
in our society were not clearly aware of the functional reason why our
restrictive sexual norms existed in the first place, they opposed any
change in these norms even after these norms ceased to be necessary.
Since many people (particularly those of a religious nature) accepted
that these norms existed simply because of the auxiliary consequences
(e.g. "it is a sin"), they saw no reason why these norms needed changing even after other suitable methods of birth control had been perfected.

The much discussed "generation gap" between the older and younger
generations in the 1960s and '70s is probably largely due to the cultural
lag in the area of sex norms. The older people accepted that these
restrictive sex norms were valid simply because the religious leaders
said so. The young people (who had always found these rules very
frustrating), realized at least on some level, that if other improved
methods of birth control were used it was no longer necessary for them
to remain celibate. This conflict between the older or "conservative"
and the young or "progressive" people arose largely because these two
groups were not talking the same language. There is really no common
ground here because they accept different justifications for these restrictive sexual rules. The older religious-conservative group undoubtedly delayed the sexual revolution considerably because they held
most of the positions of authority and power, but as the younger "progressives" gradually attain more and more leadership positions this
conservative power base is beginning to show some cracks. As with
some scientific revolutions, the sexual revolution may not be complete
until all the older conservative adherents have died.

In most developing countries, the "cultural lag" in this area appears to be much more pronounced than in the industrialized societies. Probably because Europe and America experienced a longer period of adaptation as methods of birth control were slowly improved over the past century, our cultural lag was actually relatively small. We had already made a considerable change in our norms before the pill was introduced in 1961, thus the jolt to the system was much less severe than if the pill had been discovered in the middle of the Victorian era, say in 1861. The developing countries may now be experiencing the equivalent of this. Modern methods of birth control have been thrust upon them virtually all at once, and especially since religion is still relatively powerful in many of these societies, they have been much slower to change their premarital sexual norms. Although modern methods of birth control are often available, these societies are still very restrictive of premarital sex behaviour. China, although a secular society, is a good example here. Although oral contraceptives are freely available, apparently very little sexual activity occurs in single people.

Although very little change in the sexual norms in these societies has occurred as yet, the theory proposed here predicts that this change will sooner or later occur. The cultural lag may last longer in some societies, due to numerous factors, but once a society acquires modern methods of birth control, these changes are inevitable regardless of whatever justifications are presently accepted for the existence of these norms. Thus as birth control becomes more effective, safer and readily available around the world, it seems inevitable that premarital sex will eventually become completely accepted by all societies and even all religions.

A second reason for the "cultural lag" in the area of sex norms, is concerned with the "vested interests" men have in maintaining their privileged position. Because the sex-pregnancy hypothesis implies that more stringent restrictions should be placed on the behaviour of women, men came out much better (at least in comparison to women) concerning these restrictive sex rules. Men had much more freedom in many areas than did women. Because men do not get pregnant, they were much less bound by these restrictive norms. And after they acquired this advantageous "superiority" over women, many men are now reluctant to relinquish it. The "cultural lag" in this area is undoubtedly being aided by many men's desire to maintain their relatively privileged status over women. However, this resistance to change by many men cannot be more than a delaying tactic in restoring equality of the sexes. Once a society has changed its underlying hypothesis, change is inevitable.

Remember that among the primitive permissive societies described earlier, children are free to engage in sexual activities with each other as soon as they are capable. Early sex play usually involves many forms of mutual masturbation, oral-genital contacts and usually ends in attempted copulation. In many of these societies there are no virgins after the age of ten to twelve. Thus, it seems likely that with effective birth control modern societies are headed towards a return to this "natural" (i. e. lack of restrictive prohibitions) permissive state. The mere thought of this may shock many of today's readers, but many activities which we accept as commonplace today would have shocked Victorian readers. In the future, it seems inevitable that all societies will again allow their children to engage in such sexual activities with each other whenever they wish. This permissive attitude apparently existed for almost all of our pre-history (1-4 million years) before we learned the sex-pregnancy relationship and we are now only returning to this "natural" state because we have found other ways to prevent conception. The "restrictive" era of human sexual history which is now apparently drawing to a close, may well turn out to be a relatively short (a few thousand years) period of our existence.

The acquisition of effective forms of birth control may also explain why many western societies have liberalized their abortion laws recently. Abortion (and infanticide) have undoubtedly occurred in all societies, but are often socially disapproved of (both in and out of marriage), except as a means of the last resort in limiting futher growth of a severely overpopulated society. Except under the harshest conditions, most primitive societies encourage the production of the maximum number of children. And as already discussed, most societies have decided that in order to reduce illegitimate pregnancies, celibacy in young people is a preferable method of birth control to widespread abortion in single girls.

Abortion was discouraged or even prohibited in most societies (although it often remained a common practice), not only because it was usually dangerous for the mother (whether single or married), but also because like other forms of birth control, its acceptance and use would have encouraged the single people to engage in sexual activities. And once most single people had ceased to be celibate, many single girls would have to have several abortions before they married.

An exception to this restrictive attitude occurred in the Greco-Roman world where abortion was legal. Both Judaism and early Christianity condemned abortion throughout the middle ages but severe criminal sanctions were common in Christian countries only in the nineteenth

century. The first important departure from this nineteeth century
pattern occurred in the USSR in 1920 when abortion at the request of the
mother was legalized. However, the USSR later temporarily returned
to a restrictive abortion law (1936-1955) in an attempt to increase its
population. Sweden and Denmark began to liberalize their abortion laws
in the 1930s. Britain passed a liberal abortion reform in 1967 and the
USA allowed legal abortions in 1973. Many other industrialized countries have also passed similar abortion reforms.

The two medical advances in this century which allowed this liberalization were (1) abortion became a relatively safe medical procedure
for the mother, and (2) other effective methods of birth control were
developed. Thus legalizing abortion would now not endanger the mother's life and would not precipitate a flood of abortions by both single
and married women. Modern societies have been able to sanction abortion because it is now safe for the mother, and thanks to other effective
forms of birth control, it is now necessary to use abortion only occasionally to catch a few "accidental" pregnancies (where other methods
of birth control have failed). Even where abortion has been legalized,
mothers are usually encouraged to use other forms of birth control as
the preferred alternative. Thus, the more effective the other forms of
birth control in a society, the more willing a society may be to legal-
ize abortion.

The effectiveness of the other forms of birth control as a major
factor in deciding the attitude a society adopts concerning abortion,
may help to explain why abortion was widely accepted in the Greco-
Roman culture. Remember that much of the knowledge about birth control was lost during the middle ages, thus we had to rediscover much
of this learning in the eighteeth and nineteenth centuries. The ancient
Greeks and Romans had openly allowed abortion because their knowledge of other forms of birth control was advanced sufficiently to mean
that abortion could have been used only as a backup method of control.
Both Cicero and Galen condemned indiscriminate abortion and Soranus
of Ephesus in the second century advocated prevention of conception as
preferable to repeated abortions. This indicates concern in the Greco-
Roman society not only with large numbers of abortions, but also implies that other forms of birth control were sufficiently developed to
prevent most unwanted pregnancies.

Of course, many other factors influence a society's attitude towards
abortion. One of the most important of these is whether a society considers itself underpopulated or overpopulated. Several countries have
restricted abortion, other forms of birth control and even divorce when

the leaders felt it important to increase the population as soon as possible (e.g. during or after a war or epidemic). Likewise, an overpopulated society may turn a blind eye towards restrictive rules (although often not actually changing the laws) concerning abortion, birth control and even illegitimate births. In this case, as in many others, it is clear that societies limit the individual's freedom when it is believed necessary to do so to ensure the future survival of society. The survival of society always takes precedence over the "rights" of individuals.

In modern industrialized societies, although we now have a vast array of effective methods of birth control, the rate and acceptance of illegitimate births has been increasing rather than decreasing. Although we are now perfectly able to prevent conception, the incidence of illegitimate pregnancies in many countries seems to be increasing. All societies have encouraged women to have children only within the institution of marriage. It was always felt preferable for the mother to have a man to help her provide food, clothing and shelter for the child for an extended period of time. With our modern methods of birth control, why should the rate of children born out of marriage be increasing? The answer is that science has provided us not only with effective birth control, but also a vastly improved standard of living. It is now possible for a woman to bring up a child on her own without too much difficulty. Improved economic standards, employment prospects and free state day schooling mean that it is now quite possible for one person to rear a child on his or her own. In other words, having children outside of marriage is becoming more common and more accepted because the functional reason for prohibiting illegitimate births has now largely ceased to operate. It is no longer necessary to require that children be born only within marriage because it is no longer absolutely essential for a man to help with the rearing of children. The norm against illegitimate births is being relaxed because environmental conditions have changed. The easier it is for one person to bring up a child, the less important it is for a society to prohibit illegitimate births. Although it may always be considered preferable to have two people to share in the childrearing, the necessity of instituting strong sanctions against illegitimacy has now ceased to operate in modern societies. There is also a "cultural lag" in this area similar to the lag concerning sex outside of marriage, however. Many people still believe that illegitimacy is "wrong" simply because religion says so.

As mentioned earlier, the growth of "women's liberation" (that is, the reduction of the "double standard" in norms) in this century in industrialized societies and particularly since the 1960s can largely be attributed to the changes introduced in these societies because of effective birth control, the ability of women to control their own fertility, and the increased standard of living. The introduction of effective contraceptives and a greatly increased standard of living, probably did more to "liberate" women than all the women's demonstrations and meetings held in this century. When taking the pill, a woman has virtually no more chance of becoming pregnant than does a man. Many of the old restrictions and prohibitions surrounding a woman's behaviour concerning sexual activity were now functionally unnecessary. Women were now as "free" to engage in sexual activity as men. A woman could also now *choose* to pursue a career or have children or both if she desired. There is also a considerable cultural lag in this area caused in large part by the "vested interests" men have in maintaining their advantageous position, but it seems probable that as a result of these changes, women will soon have more freedom and equality world-wide than they have had since pre-historical humans first learned the sex-pregnancy connection.

At the beginning of this chapter I quoted sections from the first part of the Judaeo-Christian bible. The reader is now requested to re-read this passage (see p. 265). You may now have some idea why I began this chapter with this quotation. The passage explains that the first humans originally lived in the "Garden of Eden", but that after they had partaken of the "tree of knowledge", their "eyes were opened" and then they "knew they were naked" and were sent forth from the garden "to till the ground".

This legend may well have been passed on by word of mouth from generation to generation for hundreds or even thousands of years before it was first written down and eventually became part of the Bible. Myths and legends may often, though distorted, adapted and symbolized to some extent by repeated re-tellings, contain some essence of an important event which occurred in our prehistory. It is possible that this legend may be telling us, in a somewhat stylized manner, of how early humans acquired some important new knowledge and what the consequences of this learning were. I do not claim that this is the one "true" interpretation of this passage, certainly there are already many different interpretations of this legend. I only claim that certain selected sections are consistent with the theory of human learning presented here and may actually suggest some interesting details which we would otherwise never realize. It is at least possible that this legend contains the only direct knowledge from our ancestors which we are ever likely to have concerning important events of our prehistory.

The first humans lived in a garden in which they could eat freely from every tree except the "tree of the knowledge of good and evil". The man and woman lived together in this garden naked and they "were not ashamed". However, the woman saw that the tree of knowledge was good "for food" and "to be desired to make one wise", so she partook of this knowledge and shared it with her husband. "And the eyes of them both were opened, and they knew that they were naked".

This legend may well be telling us about at least one and possibly two prehistoric scientific discoveries, that is when we ate of the tree of knowledge. It may be telling us what happened after our ancestors learned that sex causes pregnancy. Before early humans learned the sex-pregnancy connection, they lived together and were "not ashamed" of their nakedness. They were unaware of the consequences of sex and that sex had anything to do with "good and evil".

However, the woman saw that this knowledge was good to make one wise. Could it be that a woman made this very important scientific discovery? Certainly only females experience both sex and childbirth. Men are always interested in sexual activities, but usually much less interested than women in pregnancy and childbirth. Also, women know when they have had sex or not! Remember that the Trobriander men were misled because they believed that some very ugly women had never had sex, but nevertheless had children. These women, of course, knew better.

After humans had acquired this knowledge, their "eyes were opened" and they "knew they were naked". After we had learned the sex-pregnancy relationship, we also realized some of the important consequences of this knowledge. We became "as gods, knowing good and evil". What humans believe is moral or immoral is in large part determined by the hypotheses they accept. Once we had accepted the sex-pregnancy hypothesis, we also eventually changed our ideas about what was "good and evil". The restrictive era of human sexuality began once humans realized the long-term consequences of sexual intercourse.

This legend may also tell us about the consequences of early humans learning the seed-plant relationship. Before we partook of this knowledge we lived in a garden, gathering food from all the plants. Again it was a woman who saw that this knowledge was good "for food" and as a consequence, they left the garden "to till the ground". We ceased to be only gatherers and became producers of food. This connection may also have been first proposed by a woman. The women usually do the gathering in primitive societies and also the planting in simple agricultural societies. As I already proposed in Chapter 7, it is therefore quite possible that it was a woman who first discovered the seed-plant connection.

This biblical legend also tentatively suggests that both of these discoveries may have been made at near the same time (perhaps even by the same person?). Certainly the two elements are thoroughly mixed in the story and both discoveries have to do with reproduction—one in plants and the other in animals. This suggestion strains even the widest realms of credibility, however.

The most important point in this legend here, however, is the open implication that early humans, at least at first, realized the enormity of the consequences of gaining the knowledge of the sex-pregnancy relationship. Once we had eaten of the tree of knowledge, our eyes were opened and we knew good and evil. The suggestion is clear, that at least at first, we realized that it was the extent of our knowledge which determined our ideas about what was right and wrong. Although this realization was later forgotten or suppressed, the humans who first made this discovery were uncomfortably aware that it was the state of our own knowledge which decided what was moral or immoral. Once we knew the long-term consequences of sex, we knew when sex was "good" and when it was "evil". It is not the gods who decide what is right and wrong, it is ourselves.

Humans do indeed differ in kind from all the other animals. We are the only animals able to create our own symbolic language and the only animals able to learn separated relationships. And what is more, as a consequence of this unique learning ability, we are, at least compared to the other animals, "as gods, knowing good and evil".

Appendix I

Summary

Humans have long pondered the question of the nature of the difference which separates our species from all the other animals. Through the centuries, various proposals have been suggested to mark human intelligence as different in kind from that of all the other species. Examples include use of tools, tool making, abstract thought, language, culture, art, humour, foresight and so on.

The only one of these proposals that is still a source of argument among social scientists is language. Recent experiments in which chimpanzees were successfully taught several forms of human symbolic language have added fuel to this controversy. However, by considering all the evidence very carefully and by slightly re-defining certain terms used in this debate, it is possible to show that humans do indeed have a unique ability in this area. Humans are the only animals capable of symbolizing--that is creating our own symbols. Other animals can learn *our* symbols, but they cannot produce their own.

However, humans have an additional unique ability which scientists have yet to recognize. This ability is in the area of "learning". All animals are capable of some learning, that is changing their behaviour in light of experience. The scientific study of learning began with the work of Pavlov and Thorndike in the first part of this century. Learning psychologists have always held that all types of learning in all species must be contiguous, that is that animals are capable of learning a connection or association between two events only if these events occur close together in time. If more than a few seconds separate the events, learning is not possible. With only two recently discovered and very narrow exceptions to this principle, all learning psychologists have accepted that all learning in all species must be contiguous. No psychologist has ever proposed that Homo sapiens might be an exception to this principle of contiguity.

Scientists themselves, of course, engage in a type of learning. Learning psychologists have, of course, accepted that humans are capable of scientific learning, but they have referred to higher human learning as a "cognitive" type of learning (meaning that it occurs inside

the brain or mind). When learning psychologists studied the learning
in all the other species, they looked at it from the "outside" that is in
terms of behaviour. Psychologists always defined learning in terms of
behaviour. But when learning psychologists studied human learning,
they unconsciously switched to thinking about it from the "inside", that
is to what happened inside our brain. They thought about other animals'
learning from the "outside" (behaviour), but they still thought about
human learning from the "inside" (cognition).

Learning psychologists probably inherited this dual outlook from
Descartes (1637). Descartes' famous dualism of mind and matter still
permeates the thinking of scientists today. The "hard" or natural
sciences (e.g. physics, chemistry and biology) deal exclusively with
the objective material world, whereas the "soft" or social sciences
(psychology, philosophy, sociology) deal mainly with the subjective
internal phenomena of the "mind". When psychology first emerged as
a new science in the nineteenth century, it was universally accepted
that psychology was concerned with studying the phenomena of the mind.
However, beginning in the first part of this century, behaviouristic
psychologists made an ambitious attempt to make their sub-field a
"hard" science by considering only objective, observable phenomena
(behaviour), and deliberately ignoring any subjective internal mental
experiences. This procedure worked very well when studying learning
in other animals, but the behaviourists were still unable to shake off
Descartes' dualism when it came to their own learning. They considered other animals' learning in terms of behaviour, but they still looked
at human learning from the "inside", that is in terms of "cognition".
Our learning occurs in the "mind" (cognition), while other animals'
learning is part of the objective, material world (i.e. a change of
behaviour).

My proposal is that if only for consistency's sake, we should consider human learning in the same terms which we use when studying
the learning of all the other animals. We should consider human learning from the "outside" (in terms of behaviour). As psychologists we
should consider science from a behaviouristic point of view. As philosophers we should abandon Descartes' dualism and step outside of
ourselves and view ourselves as part of the material world. Thus my
argument is basically a philosophical one, although it is also scientific
in the sense that it is testable and falsifiable.

When science is looked at in this way, it becomes obvious that science is a non-contiguous form of learning. Scientific discoveries consist of learning a connection between events which are actually sepa-

rated by time. Humans are able to *learn* (change their behaviour) even
when the events are separated by minutes, hours, days, weeks and
even years.

More specifically then, I am proposing that learning psychologists
have long made a fundamental error in their logic. Because they considered learning from two points of view, in terms of behaviour for all
other species, but in terms of cognition for humans, they have unknowingly been accepting two ideas or principles which are, in fact, logically contradictory. They have asserted that:

(1) Learning in all species must be contiguous.

(2) Humans are capable of scientific learning.

My proposal is that these two statements are not logically compatible. Once you view science from the "outside", it becomes immediately apparent that if you accept (2), then you cannot also accept (1).
It is the principle of contiguity which is suspect.

Thus, humans have a second unique ability. We are the only animals
capable of "separated" learning, that is, learning by trial and error
or experimentation where two or more events are separated by 60
seconds or more. Humans are able to accomplish this type of learning
by "hypothesizing", that is proposing a connection between events separated by time. Thus we have not one, but two unique abilities. We
are the only animals capable of symbolizing and also the only animals
capable of hypothesizing.

What are the consequences of looking at science in this new way?
What new insights or understandings become apparent if we switch over
to considering science as a form of non-contiguous learning? There
are at least eight major new consequences of the theory as well as
numerous other implications of lesser importance. Only the major
consequences will be summarized here. Although up to this point the
argument has been mainly in the field of learning psychology, the consequences or implications of the theory are in the fields of ethology,
anthropology and the history and philosophy of science.

The first consequence is concerned with "foresight" that is, anticipating what will happen in the future as a result of experience. Although it has sometimes been suggested that the foresight of humans
is greater than that of other animals, no one has ever proposed a specific limit on other animals' foresight. However, once we accept that
all animals except humans are limited to learning when the events are
separated by not more than 60 seconds, then it becomes apparent that
their foresight is also similarly limited. Once an animal in a Skinner
box has learned the connection between pressing the lever and the drop-

ping of the food, he has some foresight about what will be likely to
happen in the future after he presses the lever. But since he is only
able to learn such a connection if the food drops a few seconds after he
presses the lever, his foresight is also limited to a few seconds into the
future. Therefore, no other animal is able to anticipate, as a result
of experience, what will be likely to happen more than one minute into
the future.

Other animals often display behaviours such as migration, hibernation, nest-building, hoarding of food and so on which seem to suggest
that they are anticipating the distant future. However, it can be shown
by experiment that in all such cases, these behaviours are instinctive,
rather than a result of experience or learning. The animals are born
with this "knowledge", they perform these activities without knowing
why they do so.

The second consequence of my theory is concerned with the history
of science. Books dealing with this subject usually start with the ancient Greeks with perhaps brief mention of the Egyptians and Babylon-
ians. The implication of this is that science has only existed for a few
thousand years, perhaps since the advent of writing, and that before
this time, early humans made few if any "scientific discoveries". For
example, Cassirer (1944, p. 207) suggested that science "is a very
late and refined product that could not develop except under special
conditions. Even the conception of science, in its specific sense, did
not exist before the times of the great Greek thinkers." However, by
looking at science from the point of view of "separated" learning, and
by incorporating archaeological evidence, it is possible to argue that,
on the contrary, science has a very long pre-history, and what is more,
in some cases, pre-historical scientific discoveries can be named,
dated and the locations in which they first occurred identified.

Anthropologists and archaeologists have learned from excavations
that early humans very gradually acquired more and more new behaviours in the thousands of years of our pre-history. For example, the
earliest humans did not even possess the use of fire, and it was only
about 10,000 years ago that humans began to plant crops. Anthropologists have proposed several different ideas as to why humans suddenly
began engaging in such activities which they had never previously practised. In the case of agriculture for example, anthropologists have
suggested that the pressure of over-population forced humans to begin
planting crops, or that after the last ice age, it became warmer and
agriculture thus became practicable and so forth. It has sometimes
even been suggested that the commencement of these activities must

have been to some extent a result of human learning. However, it has never been specifically pointed out that these advances were not the result of just any type of learning, but of the uniquely human scientific or "separated" learning. Only humans are capable of learning the connection between the placing of seeds in the ground and the appearance of small plants several *days* later. The first humans failed to engage in agriculture because they were ignorant of how to do it. Early humans could only commence such activities after a *scientific* discovery had been made. Science did not suddenly begin in the Greek civilization, rather it is as old as humanity itself.

The third main consequence of the theory deals with the problem of why these pre-historical scientific discoveries came so few and far between. If early humans were our intellectual equals, why did it take them so long? For example, why did our ancestors wait at least 98% of our existence before discovering the simple seed-plant separated relationship? If early man was our intellectual equal, how do we explain his scarcity of scientific discoveries compared with our abundance?

All human societies accept hypotheses to explain their environment. However, in today's primitive tribes and also probably early human tribes, these hypotheses were mainly of a magical or religious nature, rather than of a scientific nature. Science is based on man's unique ability to hypothesize, but magic and religion are also based on this same ability. Since early humans were surely our intellectual equals, they were just as capable of hypothesizing as we are, but because their hypotheses were mostly of a magical or religious nature, their scientific discoveries were rare.

Science is distinguished from its two cousins, magic and religion, by the requirement that scientific hypotheses must be testable or, more precisely, falsifiable before they are even considered. Science will not even consider a hypothesis unless it is possible, at least in principle, to prove the hypothesis wrong, whereas it is virtually impossible to prove any magical or religious explanation wrong.

The problem which prevented early man from making more scientific discoveries was not in the proposing of hypotheses, but rather in the accepting of the criterion for deciding which hypotheses are to be considered. Only very slowly and reluctantly did humans accept the notion that a hypothesis should be considered only if it is possible, at least in theory, to prove the hypothesis wrong! Even though magical and religious explanations prevented anxiety from arising in early man, the snag was that they also impeded the proposing of new hypotheses

and the acceptance of scientific hypotheses even after they had been
proposed and tested.

The fourth consequence considers on a deeper level my earlier suggestion that humans have *two* unique abilities, the ability to hypothesize
and the ability to symbolize. I now conclude that these two abilities
are actually two manifestations of the *same* root ability, namely the
capacity of "imagination", that is, the ability to propose relationships
between events which are separated by time. Of course it has been
suggested many times before that humans are the only animals capable
of imagination, but this term has never previously been defined in such
a way as to allow this proposal to be tested. Thus, I propose a new
integrative explanation of why humans are the only animals that have
symbolic language, representational art, magic, religion and science
(see the diagram on p. 216).

Fifth, I propose a new and specific objection to the widely accepted
view of the "cultural relativism" school of anthropology which holds
that "primitive" tribes are not really primitive. Although this school
is correct in pointing out that primitive tribes have complex and sophisticated forms of language, aesthetic art, magic and religion, there is
one area in which these tribes are obviously and demonstrably primitive and inferior to modern societies. Because of the type of hypotheses which they accept, these tribes are much less able to control and
manipulate their environment. Because they accept hypotheses which
are seldom falsifiable, they are usually unable to test and, if necessary,
reject hypotheses which are ineffective in manipulating their environment. Primitive tribes may be "in harmony" with nature, but this also
means that they are at the mercy of nature during droughts, floods,
plagues, famines and so on. They are much less able to control nature
for their own benefit and survival.

Sixth, I propose a new explanation why humans are the only animals
with rules, laws, morals and other types of norms and why these rules
differ so much from one society to another. Humans are the only animals to enforce rules or norms because we are the only animals to
anticipate the distant future. Norms are set up to encourage people to
behave in a way which that society believes will have positive survival
benefits in the distant future. All human societies have norms because
all societies are attempting to control the future. We are all attempting
to ensure that the future will turn out to be favourable for our society's
survival. Norms or rules differ drastically between societies because
societies accept drastically different hypotheses explaining their environment. Our freedom is always limited because we always have one

eye on the distant future.

Seventh, I propose a solution to the vexing problem of why some norms (particularly morals) in certain societies appear to be objectively functional but for reasons which are unknown to the members of that society. I suggest the function of these rules has been forgotten or even deliberately suppressed in these societies in an attempt to ensure their compliance.

And last, I consider the problem of sexual norms in different societies in light of the above proposals. I conclude that restrictive sexual rules were probably introduced in human societies only after humans had accepted the sex-pregnancy hypothesis. As long as we didn't know the long term consequences of sexual intercourse, we had no reason to prohibit this activity in single people. Restrictive sexual rules are the result of a scientific discovery. It is no use blaming religion for the centuries of sexual repression, religion only provided the justification for these rules. It was the acceptance of the sex-pregnancy hypothesis which necessitated the enforcement of sexual prohibitions in single people.

Likewise, the recent "sexual revolution" in modern industrialized societies is also the result of yet another scientific discovery. Science has at long last discovered new ways to prevent these undesirable future consequences of sexual activity in single people. We are now capable of accomplishing the same task with much less frustration and repression. It appears that the era of sexual repression will turn out to be a relatively brief era in human existence. The day will come in the not too distant future, when humans the world over will consider the idea of punishing children and young people for engaging in sexual activity as as strange, unnecessary and harmful as we consider the idea of burning people for engaging in "witchcraft".

Appendix II

Excerpts from

Psychology as the behaviourist views it [*]

by John B. Watson   1913

Psychology as the behaviourist views it is a purely objective experimental branch of natural science. Its theoretical goal is the prediction and control of behaviour. . . . The behaviourist, in his efforts to get a unitary scheme of animal response, recognizes no dividing line between /the study of/ man and brute. The behaviour of man, with all of its refinement and complexity, forms only a part of the behaviourist's scheme of investigation.

It has been maintained by its followers generally that psychology is a study of the science of the phenomena of consciousness. . . . The world of physical objects (stimuli, including here anything which may excite activity in a receptor), which forms the total phenomena of the natural scientist, is looked upon merely as means to an end. . . . It is agreed that introspection is the method *par excellence* by means of which mental states may be manipulated for purposes of psychology. On this assumption, behaviour data (including under this term everything which goes under the name of comparative psychology) have no value *per se*. They possess significance only in so far as they may throw light upon conscious states. Such data must have at least an analogical or indirect reference to belong to the realm of psychology.

Indeed, at times, one finds psychologists who are sceptical of even this analogical reference. Such scepticism is often shown by the question which is put to the student of behaviour, "what is the bearing of animal work upon human psychology?". . . We must frankly admit that the facts so important to us which we have been able to glean from extended work upon the senses of animals by the behaviour method have contributed only in a fragmentary way to the general theory of human sense organ processes, nor have they suggested new points of experimental attack. The enormous number of experiments which we have carried out upon learning have likewise contributed little to human psychology. It seems reasonably clear that some kind of compromise

---

[*] Reprinted from 'Psychological Review', Vol. 20, 1913, p. 158-177.

must be effected: either psychology must change its viewpoint so as to
take in facts of behaviour, whether or not they have bearings upon the
problems of 'consciousness'; or else behaviour must stand alone as a
wholly separate and independent science. Should human psychologists
fail to look with favour upon our overtures and refuse to modify their
position, the behaviourists will be driven to using human beings as
subjects and to employ methods of investigation which are exactly comparable to those now employed in the animal work. . . .

One can assume either the presence or the absence of consciousness
anywhere in the phylogenetic scale without affecting the problems of
behaviour by one jot or one tittle; and without influencing in any way
the mode of experimental attack upon them. . . .

I do not wish unduly to criticize psychology. It has failed signally,
I believe, during the fifty-odd years of its existence as an experimental
discipline to make its place in the world as an undisputed natural science. Psychology, as it is generally thought of, has something esoteric
in its methods. If you fail to reproduce my findings, it is not due to
some fault in your apparatus or in the control of your stimulus, but it
is due to the fact that your introspection is untrained. . . .

The time seems to have come when psychology must discard all
reference to consciousness; when it need no longer delude itself into
thinking that it is making mental states the object of observation. We
have become so enmeshed in speculative questions concerning the elements of mind, the nature of conscious content (for example, image-
less thought, attitudes, and Bewusseinslage, etc.) that I, as an experimental student, feel that something is wrong with our premises and the
types of problems which develop from them. There is no longer any
guarantee that we all mean the same thing when we use the terms now
current in psychology. . . . I firmly believe that two hundred years
from now, unless the introspective method is discarded, psychology
will still be divided on the question as to whether auditory sensations
have the quality of 'extension,' whether intensity is an attribute which
can be applied to colour, whether there is a difference in 'texture"
between image and sensation and upon many hundreds of others of like
character. . . .

I was greatly surprised some time ago when I opened Pillsbury's
book and saw psychology defined as the 'science of behaviour'. A still
more recent text states that psychology is the 'science of mental behaviour'. When I saw these promising statements I thought, now surely
we will have texts based upon different lines. After a few pages the
science of behaviour is dropped and one finds the conventional treat-

ment of sensation, perception, imagery, etc., along with certain shifts
in emphasis and additonal facts which serve to give the author's personal imprint. . . .

I believe we can write a psychology, define it as Pillsbury, and
never go back upon our definition: never use the terms consciousness,
mental states, mind, content, introspectively verifiable, imagery,
and the like. I believe that we can do it in a few years without running
into the absurd terminology of Beer, Bethe, Von Uexkull, Nuel, and
that of the so-called objective schools generally. It can be done in
terms of stimulus and response, in terms of habit formation, habit
integrations and the like. Furthermore, I believe that it is really
worth while to make this attempt now.

The psychology which I should attempt to build up would take as a
starting point, first, the observable fact that organisms, man and
animal alike do adjust themselves to their environment by means of
hereditary and habit equipments. These adjustments may be very adequate or they may be so inadequate that the organism barely maintains its existence; secondly, that certain stimuli lead the organisms
to make the responses. In a system of psychology completely worked
out, given the response the stimuli can be predicted; given the stimuli
the response can be predicted. Such a set of statements is crass and
raw in the extreme, as all such generalizations must be. Yet they are
hardly more raw and less realizable than the ones which appear in the
psychology texts of the day. . . .

In experimental pedagogy especially one can see the desirability of
keeping all of the results on a purely objective plane. If this is done,
work there on the human being will be comparable directly with the
work upon animals. . . . We need to have similar experiments made
upon man, but we care as little about his 'conscious processes' during
the conduct of the experiment as we care about such processes in the
rats. . . . The man and the animal should be placed as nearly as possible under the same experimental conditions. . . .

The situation is somewhat different when we come to a study of the
more complex forms of behaviour, such as imagination, judgment,
reasoning, and conception. At present the only statements we have of
them are in content terms. Our minds have been so warped by the
fifty-odd years which have been devoted to the study of states of consciousness that we can envisage these problems only in one way. We
should meet the situation squarely and say that we are not able to carry
forward investigations along all of these lines by the behaviour methods
which are in use at the present time. . . .

In concluding, I suppose I must confess to a deep bias on these questions. I have devoted nearly twelve years to experimentation on animals. It is natural that such a one should drift into a theoretical position which is in harmony with his experimental work. . . . Certainly the position I advocate is weak enough at present and can be attacked from many standpoints. Yet when all this is admitted I still feel that the considerations which I have urged should have a wide influence upon the type of psychology which is to be developed in the future. What we need to do is to start work upon psychology, making *behaviour*, not *consciousness*, the objective point of our attack. Certainly there are enough problems in the control of behaviour to keep us all working many lifetimes without ever allowing us time to think of consciousness *an sich*. Once launched in the undertaking, we will find ourselves in a short time as far divorced from an introspective psychology as the psychology of the present time is divorced from faculty psychology.

SUMMARY

Human psychology has failed to make good its claim as a natural science. Due to a mistaken notion that its fields of facts are conscious phenomena and that introspection is the only direct method ascertaining these facts, it has enmeshed itself in a series of speculative questions which, while fundamental to its present tenets, are not open to experimental treatment. In the pursuit of answers to these questions, it has become further and further divorced from contact with problems which vitally concern human interest.

Psychology, as the behaviourist views it, is a purely objective, experimental branch of natural science which needs introspection as little as do the sciences of chemistry and physics. It is granted that the behaviour of animals can be investigated without appeal to consciousness. . . . The position is taken here that the behaviour of man and the behaviour of animals must be considered on the same plane; as being equally essential to a general understanding of behaviour. It can dispense with consciousness in a psychological sense. . . .

This suggested elimination of states of consciousness as proper objects of investigation in themselves will remove the barrier from psychology which exists between it and the other sciences. . . .

Psychology as behaviour will, after all, have to neglect but few of the really essential problems with which psychology as an introspective science now concerns itself. In all probability even this residue of problems may be phrased in such a way that refined methods in behaviour (which certainly must come) will lead to their solution.

References

PREFACE
Skinner, B. F. Are Theories of Learning Necessary? 'Psychological Review', 57, 193-216, 1950.

CHAPTER 1
Aristotle, 'Historia Animalium', London: G. Bell, 1902.
_____, 'Aristotle De Anima', R. D. Hicks, (Ed.) Cambridge: University Press, 1907 (p. 134-135).
Beach, Harlan P. 'Geography and Atlas of Protestant Missions', New York: Student Volunteer Movement for Foreign Missions, 1901.
Beals, R. L. & Hoijer, H. 'An Introduction to Anthropology', New York: Macmillan, 1959.
Bitterman, M. E. Phyletic Differences in Learning, 'American Psychologist', 20 (1965) 396-410.
Dart, Raymond A. 'Adventures with the Missing Link', London: Hamish Hamilton, 1959.
Darwin, Charles 'The Descent of Man and Selection in Relation to Sex', New York: Collier & Son, 1871 (Chapters 3 & 18).
Descartes, Rene 'Discourse on Method', 1637.
Gallup, Gordon G. Jr. It's Done with Mirrors--Chimps and Self-Concept, 'Psychology Today', 1971, 4(10), 58-61.
Goldstein, K. and Scheerer, M. Abstract and concrete behaviour: an experimental study with special tests. 'Psychol. Monogr'. 1941, 53(2).
Goldstein, K. 'After-effects of brain injuries in war', New York: Grume and Stratton, 1942.
Hanfmann, Euginia, Rickers-Ovsiankina, Maria, and Goldstein, K. Case Lanuti: Extreme concretization of behaviour due to damage of the brain cortex. 'Psychol. Monogr' . 1944, 57(4).
Kawantura, Syunzo The Process of Sub-culture Propagation among Japanese Macaques, 'Journal of Primatology', Vol 2, No. 1, 1959, p. 43-60 (English edition published in March, 1962).
Khroustov, H. F. Formation and Highest Frontier of the Implemental Activity of Anthropoids. Paper delivered to the Seventh Internation Congress of Anthropological and Ethological Sciences in Moscow 1964.
Kohler, W. 'The Mentality of Apes', London: Routledge & Kegan Paul 1921.

Lawick-Goodall, Jane van 'In the Shadow of Man', London: Collins, 1971.

Lloyd, A. B. Acholi country: Part II 'Uganda Notes 5', 18-22, 1901.

Locke, John 'An Essay Concerning Human Understanding', London, 1706 (Book II, Chapter 11).

McDougall, William 'An Outline of Psychology', London: Methuen, 1928.

McDougall, K. D. & McDougall, W. Insight and Foresight in Various Animals Monkeys, Racoon, Rat and Wasp. 'Journal of Comp. Psychol'. Vol. 11, 237-273, 1931.

Morris, Desmond 'The Biology of Art', London: Methuen 1962.

Munn, N. L. 'The Evolution and Growth of Human Behaviour', London: Harrap, 1955.

Pikas, A. 'Abstraction and concept formation', Cambridge: Harvard Univ. Press, 1966.

Rose, Steven 'The Conscious Brain', London: Weidenfeld & Nicolson, 1973.

Schultz, Adolph H. Characters Common to Higher Primates and Characters Specific for Man, 'The Quarterly Review of Biology', 1936, 2 (3 & 4).

Thorndike, E. L. Animal Intelligence: An experimental study of the associative processes in animals. 'Psychol. Rev. Monogr. Supp. II', No. 4, 1898.

_____ 'Human Learning', Cambridge: MIT Press, 1931.

Thorpe, W. H. 'Learning and Instinct in Animals', London: Methuen, 1963.

White, Leslie A. 'The Science of Culture', New York: Farrar, Straus, and Cudahy, 1949.

Wundt, Wilhelm 'Logik. Eine Untersuchung der Prinzipien der Erkenntnis', (Vol. I: 1893; Vol. II: 1894). Stuttgart: Ferdinard Enke (English translation of quotation from Pikas, 1966, p. 10).

Yerkes, Robert M. 'Chimpanzees: A Laboratory Colony', New Haven: Yale Univ. Press, 1943.

CHAPTER 2

Cassirer, Ernst 'An Essay on Man', New Haven: Yale University Press, 1944.

Chomsky, N. 'Aspects of the Theory of Syntax', Cambridge, Mass.: MIT Press, 1965.

_____ 'Language and Mind', New York: Harcourt Brace, 1968.

Crawford, M. P. The cooperative solving of problems by young chimpanzees, 'Comp. Psychol. Monogr.', 14, (2) pp 88, 1937.

Fouts, Roger S. Use of Guidance in Teaching Sign Language to a Chimpanzee, 'Journal of Comparative and Physiological Psychology', Vol. 80, No. 3, 515-522, September 1972.

Gardner, R.A. & Gardner, B.T. Teaching Sign Language to a Chimpanzee, 'Science', Vol. 165, 664-672, August 15, 1969.

Goody, Jack & Watt, Ian The Consequences of Literacy, 'Comparative Studies in Society and History', 5, (3), 304-345, 1963.

Hayes, K.J. & Hayes, C. The Intellectual Development of a Home-Raised Chimpanzee, 'Proceedings of the American Philosphical Society', 1951, 95, 105-109.

Langer, Susanne K. 'Mind: An Essay on Human Feeling Vol II', Baltimore: John Hopkins University Press, 1972.

Premack, David The Education of Sarah: A chimp Learns the Language, 'Psychology Today', 4(4), 54-58, September, 1970.

Premack, David 'Intelligence in Ape and Man' New York: Halsted Press, 1977.

Scobie, William A Language just for two, 'The Observer Magazine', 20 May 1979, 67-71.

Seyfarth, Robert M., Cheney, Dorothy L. & Marler, Peter Monkey Responses to Three Different Alarm Calls: Evidence of Predator Classification and Semantic Communication, 'Science' (210) 14 Nov. 1980, 801-803.

Terrace, H.S., Petitto, L.A., Sanders, R.J. & Bever, T.G. Can an Ape Create a Sentence? 'Science' (206) 23 Nov. 1979, 891-902.

Thorpe, W.H. 'Bird Song: the biology of vocal communication and expression in birds', Cambridge University Press, 1961.

von Frisch, Karl Dialects in the Language of the Bees, 'Scientific American', August 1962.

White, Leslie A. 'The Science of Culture', New York: Farrar, Straus and Cudahy, Inc. 1949.

White, Leslie A. & Dillingham, Beth 'The Concept of Culture', Minneapolis: Burgess Pub., 1973.

CHAPTER 3

Carr, Harvey The Interpretation of the Animal mind, 'Psychological Review', 1927, 86-106.

Dickinson, A. & MacKintosh, N.J. Classical Conditioning in Animals. 'Ann. Rev. Psychol.', 1978, 29, 587-612.

Guthrie, E.R. 'The psychology of Learning', Revised. New York: Harper and Row, 1952.

Humphrey, G. 'The Nature of Learning in its Relation to the Living
System', London, 1933.

Hunter, W.S. The Delayed Reaction in Animals and Children, 'Behaviour Monographs', 1913, No. 2.

Kohler, W. 'The Mentality of Apes', London: Routledge & Kegan Paul,
1921.

Lett, B.T. Delayed reward learning: Disproof of the traditional theory.
'Learning and Motivation', 1973, 1, 237-246.

Lett, B.T. Visual discrimination learning with a 1-min. delay of
reward. 'Learning and Motivation,' 1974, 5, 174-181.

Lett, B.T. Long delay learning in the T-maze. 'Learning and Motivation', 1975, 6, 80-90.

Lieberman, D.A., McIntosh, D.C. & Thomas, G.V. Learning when
Reward is Delayed: A Marking Hypothesis. 'Journal of Experimental
Psychology: Animal Behaviour Process', 1979, 5, 224-242.

Lieberman, D.A. Marking: A test of two interpretations. Paper presented at the Eastern Psychological Association convention,
Hartford, Connecticut, April, 1980.

Neisser, U. 'Cognitive Psychology', New York: Appleton-Century-
Crofts, 1967.

_____ 'Cognition & Reality', San Francisco: W.H. Freeman, 1976.

Pavlov, I.P. Experimental psychology and psychopathology in animals,
1903. In Pavlov 'Lectures on conditioned reflexes', Translated by
W.H. Gantt. New York: International pp 47-69, 1928.

Perin, C.T. A quantitative investigation of the delay-of-reinforcement
gradient, 'Journal of Exper. Psychology', 1943, 32, 37-51.

Renner, K.E. Delay of Reinforcement: A historical review, 'Psychological Bulletin', 1964, 61, (5), 341-361.

Rescorla, R.A. Probability of shock in the presence and absence of
CS in fear conditioning. 'Journal of Comparative and Physiological
Psychology', 1968, 66, 1-5.

Rescorla, R.A. Pavlovian excitatory and inhibitory conditioning. In
W.K. Estes (Ed.) 'Handbook of Learning and Cognitive Processes'
Vol. 2. Hillsdale, N.J.: Lawrence Erlbaum Associates, 1975,7-36.

Revusky, S.H. The Role of interference in Association over a delay.
In W.K. Honig & P.H.R. James (Eds.) 'Animal Memory', New
York: Academic Press, p. 156-213, 1971.

Revusky, S.H. Learning as a general process with an emphasis on data
from feeding experiments. In N.W. Milgram, L. Krames & T.M.
Alloway (Eds.) 'Food Aversion Learning', New York: Plenum Press,
1977, 1-51.

Roberts, W. A. Failure to Replicate Visual Discrimination Learning
    with a 1-min. Delay of Reward. 'Learning and Motivation', 1976,
    313-325.
Skinner, B. F. 'The Behaviour of Organisms: an experimental analysis',
    New York: Appleton-Century-Crofts, 1938.
Spence, K. W. The role of secondary reinforcement in delayed reward
    learning. 'Psychol. Rev.', 1947, 54, 1-8.
_____ 'Behaviour theory and conditioning', New Haven; Yale
    University Press, 1956.
Thorndike, E. L. Animal Intelligence: An experimental study of the
    associative processes in animals, 'Psychol. Rev. Monogr. Supp II'
    No. 4, 1898.
_____ 'Human Learning', Cambridge: MIT Press, 1931.
Thorpe, W. H. 'Learning and Instinct in Animals', London:Methuen 1963.
Tolman, E. C. 'Purposive behaviour in animals and men', New York:
    Appleton-Century-Crofts, 1932.
Watson, John B. Psychology as the behaviourist views it, 'Psychol-
    ogical Review', Vol. 20, 1913, 158-177.

CHAPTER 4
Perin, C. T. A quantitative investigation of the delay-of-reinforcement
    gradient. 'Journal of Exper. Psychology', 1943, 32, 37-51.
Semmelweis, Ignas P. 'Die Aetiologie, der Begriff und die Prophylaxis
    des Kindbettfiebers'. Pest, Wien und Leipzig, 1861. (Complete
    English translation by F. P. Murphy in 'Medical Classics', 1941,
    5, 330-773).
Sinclair, W. J. 'Semmelweis: His life and his Doctrine', Manchester
    University Press, 1909.
Tolman, E. L. 'Purposive behaviour in Animals and Men', New York:
    Appleton-Century-Crofts, 1932.

CHAPTER 5
Hempel, Carl G. 'Philosophy of Natural Science', Englewood Cliffs:
    Prentice-Hall, 1966.
Hume, David 'Inquiry Concerning Human Understanding', London, 1739.
Kelly, Howard A. 'Walter Reed and Yellow Fever', New York: McClure,
    Phillips and Co. 1906.
Krechevsky, I. "Hypotheses" in rats. 'Psychol. Rev.' 39, 1932, 516-32.
Kuhn, Thomas S. 'The Structure of Scientific Revolutions' (2nd Ed.)
    University of Chicago Press, 1970.
Oppolzer, T. R. von 'Canon of Eclipses', New York: Dover, 1962.

Popper, K. 'Conjectures and Refutation: the Growth of Scientific Knowledge', London: Routledge and Kegan Paul, 1963.

Revusky, S. H. The role of interference in Association over a delay. In W. K. Honig & P. H. R. James (Eds.) 'Animal Memory', New York: Academic Press, p. 156-213, 1971.

Spence, K. W. 'Behaviour theory and Conditioning', New Haven: Yale University Press, 1956.

Thorpe, W. H. 'Learning and Instinct in Animals', London: Methuen, 1963.

CHAPTER 6

Barnett, S. A. 'The Rat: A Study in behaviour', London: Methuen, 1963.

Darwin, Charles 'On the Origin of Species', London: 1859.

Davis, Clara M. Self-selection of food by children, 'American Journal of Nursing', 1935, XXXV, No. 5, 403-410.

Etscorn F. & Stephens, R. Establishment of conditioned taste aversions with a 24-hour CS-US interval. 'Physiological Psychology', 1973, 251-253.

Garcia, J. , Kimeldorf, D. J. & Koelling, R. A. Conditioned aversion to saccharin resulting from exposure to gamma radiation. 'Science', 1955, 122, 157-158.

Harlow, H. F. , & Harlow, M. K. In Schrier, A. M. et al, (ed.), 'Behaviour of Nonhuman Primates', Vol. 2. London: Academic Press "The Affectional Systems" 1965.

Kalat, J. , & Rozin, P. You can lead a rat to poison but you can't make him think. In Seligman & Hager (1972) p. 115-122.

Lashley, K. S. Experimental analysis of instinctive behaviour. 'Psychol. Rev. ' 1938, 45: 445-471.

Logue, A. W. Taste Aversion and the Generality of the Laws of Learning. 'Psychological Bulletin', 1979, 86, (2) 276-296.

Lorenz, Konrad A contribution to the comparative sociology of colonial-nesting birds. 'Proc. VIII Int. Orn. Congr'. 206-218, 1938.

_____ 'Studies in Animal and Human Behaviour'. London: Methuen Vol. I: 1970, Vol. II: 1971.

Malinowski, Bronislaw 'The Sexual Life of Savages in North-Western Melanesia'. London: Routledge and Kegan Paul, 1929.

Pitt, F. 'The Intelligence of Animals: Studies in Comparative Psychology'. London, 1931.

Richter, C. P. Total self-regulatory functions in animals and human beings. 'Harvey Lecture Series', 1943, 38, 63-103.

Rozin, P. & Kalat, J. W. Specific Hungers and poison avoidance as
  adaptive specialization of learning. 'Psychological Review', 1971,
  78, (6), 459-486.
Seligman, M. E. P. , & Hager, J. 'The Biological Boundaries of Learning'. New York: Appleton-Century-Crofts, 1972.
Tinbergen, N. Specialists in nest-building. 'Country Life', 30 Jan.
  1953, 270-271.
Wilcoxin, H. C. , Dragoin, W. B. , & Kral, P. A. Illness-induced aversions in rat and quail: Relative salience of visual and gustatory
  cues. 'Science', 1971, 171, 826-828.

CHAPTER 7
Descartes, Rene 'Les traités de l'homme et de la formation du foetus',
  Amsterdam, 1680.
Farrington, B. 'Science in Antiquity', Oxford University Press, 1969.
Leach, E. R. Virgin birth, 'Proceedings of the Royal Anthropological
  Institute', 1966, 39-49.
Malinowski, Bronislaw 'The Sexual Life of Savages in North-Western
  Melanesia', London: Routledge and Kegan Paul, 1929.
Marshall, Don S. & Suggs, Robert C. (Eds.) 'Human Sexual Behaviour:
  Variations in the Ethnographic Spectrum'. Englewood Cliffs:
  Prentice Hall, 1971.
Montagu, Ashley 'Coming into Being Among the Australian Aborigines:
  A study of the Procreative Beliefs of the Native Tribes of Australia',
  London: Routledge & Kegan Paul, 2nd ed. 1974.
Nagy, Maria The child's theories concerning death. 'J. genet. Psy-
  chol.' 1948, 73, 3-27.
Sauer, C. O. 'The Domestication of Animals and Food stuffs'.
  Cambridge Mass: MIT Press, 1952.
Spiro, Melford E. Virgin birth, parthenogenesis and physiological
  paternity: an essay in cultural interpretation, 'Man' (N. S.) 1968,
  3, 212-261.

CHAPTER 8
Barnes, Barry 'Scientific Knowledge and Sociological Theory'. London:
  Routledge & Kegan Paul, 1974.
Durkheim, E. 'The Elementary Forms of the Religious Life', 1915.
Evans-Pritchard, E. E. The Intellectualist (English) Interpretation of
  Magic, 'Bulletin of the Faculty of Arts', Egyptian University (Cairo),
  vol. i, 1933.

Evans-Pritchard, E.E. Levy-Bruhl's Theory of Primitive Mentality,
    'Bulletin of the Faculty of Arts', Egyptian University (Cairo),
    vol. ii, 1934.
_____ 'Witchcraft, Oracles and Magic among the
    Azande', London: Oxford Univ. Press, 1937 (Abridged Ed. pub. 1976).
_____ 'Theories of Primitive Religion', London:
    Oxford Univ. Press, 1965.
Frazer, James G. 'The Golden Bough'. One vol. abridged ed. New
    York: Macmillan Co., 1922 (First pub. in 1890).
Gellner, E. Concepts and Society as reprinted in B.R. Wilson (ed.),
    'Rationality', 1970, Oxford: Blackwell. (First pub. in 1962).
Goody, Jack & Watt Ian The Consequences of Literacy, 'Comparative
    Studies in Society and History', 1963, 5, (3), 304-345.
Horton, Robin African Traditional Thought and Western Science, 'Africa'
    XXXVII, Nos. 1 & 2 (Jan. & April 1967), pp. 50-71 and 155-187.
Howells, W.W. 'The Heathens, Primitive Man and His Religions',
    New York: Doubleday and Co. 1948.
Levy-Bruhl, Lucien 'Primitive Mentality', 1923.
Malinowski, Bronislaw 'Magic, Science and Religion', New York:
    Doubleday and Co. 1948.
Nance, John 'The Gentle Tasaday', London: Gollancz, 1975.
Popper, Karl R. 'Logik der Forschung', Vienna, 1934. (English trans-
    lation, 'The Logic of Scientific Discovery', 1959 London).
Tylor, Edward B. 'Primitive Culture', 1871.
Wilson, Bryan R. (ed.) 'Rationality', Oxford: Blackwell, 1970.

CHAPTER 9
Bronowski, J. 'The Identity of Man', London: Heinemann, 1965.
Day, Clarence 'The Story of Yale University Press', New Haven: Yale
    University Press, 1920.

CHAPTER 10
Firth, R. 'Man and Culture: An Evaluation of the Work of Bronislaw
    Malinowski', London: Routlege & Kegan Paul, 1957.
Harris, M. 'Cows, Pigs, Wars and Witches: The Riddles of Culture'.
    New York: Vintage Books, 1974.
Hoebel, E.A. 'The Law of Primitive Man', New York: Atheneum 1954.
Kuo, Z.Y. The Genesis of the cat's behaviour towards the rat, 'Jour-
    nal of Comp. Psychol.', 1930, 11, 1-35.
Malinowski, Bronislaw 'A Scientific Theory of Culture and other Essays',
    Oxford University Press, 1944.

Ogburn, F. W. 'Social Change with Respect to Culture and Original
    Nature', Gloucester, Mass: Peter Smith, 1950.
_____ Cultural Lag as Theory, 'Sociology and Social Research',
    1957, 41,167-174.
Sherif, M. 'The Psychology of Social Norms'. New York: Harper and
    Row, 1936.
_____ Experiments in Group Conflict, 'Scientific American',
    November, 1956.
Sumner, W. G. 'Folkways: A Study of the Sociological Importance of
    Usages, Manners, Customs, Mores and Morals'. Boston: Ginn 1907.
Webb, W. P. 'The Great Frontier', Austin: Univ. of Texas Press, 1951.
White, Leslie A. 'The Science of Culture', New York: Farrar, Straus,
    and Cudahy, Inc., 1949.

CHAPTER 11
Draper, E. 'Birth Control in the Modern World', London: Allen &
    Unwin, 1965.
Ford, C. S. & Beach, F. A. 'Patterns of Sexual Behaviour', New York:
    Harper & Row, 1951.
Himes, Norman E. 'Medical History of Contraception', New York:
    Gamut Press, 1963.
Malinowski, Bronislaw 'The Sexual Life of Savages in North-Western
    Melanesia', London: Routledge & Kegan Paul, 1929.
Marshall, D. S. & Suggs, R. C. (Eds.), 'Human Sexual Behaviour:
    Variations in the Ethnographic Spectrum'. Englewood Cliffs:
    Prentice-Hall, 1971.
Montagu, A. 'Coming into Being Among the Australian Aborigines: A
    study of the Procreative Beliefs of the Native Tribes of Australia'.
    London: Routledge & Kegan Paul, 2nd ed. 1974.
Terman, L. M. 'Psychological Factors in Marital Happiness'. New
    York: McGraw-Hill Book Co. 1938.

SUMMARY
Cassirer, Ernst 'An Essay on Man', New Haven: Yale Univ. Press,
    1944.
Descartes, Rene 'Discourse on Method', 1637.

Index

Credits

Permission to quote excerpts and reproduce illustrations has been given by the following publishers: Excerpts from 'The Mentality of Apes' by Wolfgang Kohler, 1927, reprinted by permission of Routledge & Kegan Paul Ltd. Excerpt from 'The Identity of Man' by J. Bronowski copyright © 1965 by J. Bronowski, reprinted by permission of Doubleday & Company, Inc. Excerpt from 'The Story of Yale University Press' by Clarence Day, copyright 1920, reprinted by permission of Yale University Press. Excerpt from 'Psychological Factors in Marital Happiness' by L. M. Terman 1938, reprinted by permission of McGraw-Hill Book Co. The drawing on p. 22 is from 'The Conscious Brain' by Steven Rose © 1973 reprinted by permission of Weidenfeld & Nicolson Ltd. The drawing on p. 33 (by a chimpanzee) is from 'The Biology of Art' by Desmond Morris © 1962 reprinted by permission of Desmond Morris. The drawing on p. 44 is from Animal Communication by Edward O. Wilson copyright © 1972 by 'Scientific American Inc.' All rights reserved. The photograph on p. 47 is from 'Chimpanzees: A Laboratory Colony' by Robert M. Yerkes 1943, reprinted by permission of Yale University Press. The drawing on p. 73 is from 'Animal Intelligence' by E. L. Thorndike reprinted by permission of Macmillan Publishing Co. Inc. The drawing on p. 70 is from Yerkes, R. M. & Morgulis, S. The method of Pavlov in animal psychology, 'Psychologial Bulletin', 1909, 6, 257-273. The drawing on p. 75 is from 'The Behaviour of Organisms: An Experimental Analysis', by B. F. Skinner © 1938, renewed 1966, p. 49, reprinted by permission of Prentice-Hall, Inc., Englewood Cliffs, N. J. The cartoon on p. 86 is from Dennis the Menace Cartoon 1973 Used by permission of Hank Ketcham.

Send for a free list of books published by Prytaneum Press.
You may order our books from your local bookshop or you may order directly from us. Simply list (1) the title(s) of the book(s) you want (2) the price (3) your name and address. Enclose a cheque, postal or money order for the total purchase price of your book(s).

Prytaneum Press
121 Bouverie Road
London, N16 OAA
England